Advances in Atom and Single Molecule Machines

Advances in Atom and Single Molecule Machines is the first comprehensive series of books dealing specifically with single atom and molecule machines. Derived from a number of long-term European Commission Future and Emerging Technologies (FET) projects including AtMol, Elfos, Focus, Diamant, Artist, PAMS and MEMO, volumes in this series comprise topical reviews, lecture-course derived textbooks and re-worked proceedings of workshops.

Xavier Bouju · Christian Joachim
Editors

Crossroad of Maxwell Demon

Springer

Editors
Xavier Bouju
Nanosciences group
CEMES-CNRS
Toulouse, France

Christian Joachim
GNS PicoLab
CEMES-CNRS
Toulouse, France

ISSN 2193-9691 ISSN 2193-9705 (electronic)
Advances in Atom and Single Molecule Machines
ISBN 978-3-031-57903-5 ISBN 978-3-031-57904-2 (eBook)
https://doi.org/10.1007/978-3-031-57904-2

This Springer imprint is published by the registered company Springer Nature Switzerland AG
The registered company address is: Gewerbestrasse 11, 6330 Cham, Switzerland

Paper in this product is recyclable.

Preface

Thermodynamics at the nanometer and now at the picometer scale is accessible to experimental investigations thanks to the technical progress in handling and controlling one atom or one molecule at a time on a surface or in a cold trap. In chasing the Maxwell demon at the nanoscale, we have to remember all the trials of certifying or declassifying the Demon from its birth at the middle end of the nineteenth century. Mechanical, electronics, or thermodynamics machineries have been called all over the years to explore how work can be extracted from solely a single reservoir. This quest was recently extended to the quantum state space using mesoscopic qubits or single atom qubits to benefit from the quantum superposition principle in order to deviate from equilibrium. It is expected that reaching the fabrication of nano/micro-mechanical systems, counterpart of electronics machinery and of the quantum qubits, will open a new way of creating a demon.

This book presents different paths for approaching the demon from the bottom up to the mesoscale. It results from the ESIM workshop "Crossroad of Maxwell Demon" held in Toulouse (France) on the 1st and 2nd of March 2023. ESiM (*Energy Storage in Molecule(s)*) is a European HORIZON-EIC-2021 Pathfinder Open project (2022–2026) with the objective to address the pressing challenges of scalable and clean energy storage starting from the bottom (single molecule-machines). Twenty invited speakers from Europe have presented their works. To our knowledge, it was the first time that interdisciplinary sciences communities met together for a long to share their viewpoints on topics like intramolecular or quantum thermodynamics and single molecule motive power. The participants debated and exchanged in a relaxed atmosphere because tracking a demon is always enjoyable. Toulouse was also an opportune place to be for this workshop because this city has already acquired a lot of knowledge about demons. The 20 April 1230, the Dominican Inquisition was officially created in Toulouse under Pope Gregoire IX. Thomas Aquinas (St Thomas, 1225–1274) is buried in Toulouse: He developed an innovative philosophy at this time by highlighting the role of reason in accessing certain truths as a counterweight to faith alone. Many of the invited speakers took the opportunity of the workshop to present their best demon drawings almost like the illuminations of the thirteenth century!

The present volume is organized as follows. The first contribution of J. M. R. Parrondo et al. is a mini-review and addresses the information exchange between reservoirs considered as bipartite systems. G. Zicari et al. tackle the entropy production with open quantum systems in terms of a phase-space formulation. Then, G. Schaller wonders how small Maxwell's demon could be with an emphasis on autonomous electronic feedback models. M. Consiglio and T. J. G. Apollaro explain the problem of preparing a quantum many-body thermal state via a variational quantum algorithm from the perspective of quantum computers. Another example of quantum information engines is provided by M. Bowen and M. Lamblin with an emphasis on the novel concept of quantum spintronic engine. The next contribution is from S. Dago *et al.* and concerns a specific demon using an underdamped micromechanical oscillator within the stochastic thermodynamics framework. Finally, with the last three contributions (J.-P. Launay, R. Robles and N. Lorente, K. H. Au-Yeung et al.), we called on board during this workshop how the micro-reversibility principle applies inside a single molecule for its mechanics and how ultimately to tackle the Maxwell demon using a well-designed single molecule.

We expect that this new volume of the series concerning nano-thermodynamics and its possible deviation using advanced microscopic machinery will encourage the readers to enter the exploration of new practical devices from the mesoscale down to the atomic scale. The ultimate goal could be to pick up a bit of energy from a single reservoir to produce work or at least to store this energy for future usage. This is also a good opportunity to explore the mixing between quantum and semi-classical behaviors of nanoscale machinery.

We thank the European Commission through the European HORIZON-EIC program for the ESiM financial support. A very special thanks to the CEMES-CNRS and TU-Dresden personnel for their contribution to the organization of this interdisciplinary workshop.

Toulouse, France
December 2023

Xavier Bouju
Christian Joachim

Contents

Information Flows in Nanomachines

Juan M. R. Parrondo, Jorge Tabanera-Bravo, Federico Fedele, and Natalia Ares

Abstract Nanomachines are often interpreted as autonomous Maxwell's demons or information engines. This interpretation is appealing from a conceptual viewpoint but sometimes lacks a rigorous and precise formulation. Moreover, it is unclear whether this interpretation is truly useful for analyzing the performance of nanomachines or designing more efficient ones. The concepts of mutual information and information flows provide a quantitative and rigorous account of the role of information in physical systems. In this chapter, we present a pedagogical review of these concepts and apply them to a specific energy transducer at the nanoscale, which couples the tunnel transport of electrons through a carbon nanotube with its mechanical oscillations.

1 Introduction

In 1867, Maxwell introduced his celebrated demon as a Gedanken experiment, in which a 'neat-fingered being' capable of observing and operating the microscopic world could defy the second law of thermodynamics and, for instance, extract work from a single thermal bath [1]. Single-molecule machines, designed to move and/or perform work at the nanoscale, instantly bring the Maxwell demon to mind.

In Maxwell's original idea, the demon observes the velocity and position of molecules in two gases separated by a wall. Based on these observations, the demon selectively opens and closes a gate in the wall. Specifically, the demon allows fast

J. M. R. Parrondo (✉) · J. Tabanera-Bravo
Departamento de Estructura de la Materia, Física Térmica y Electrónica, and GISC, Universidad Complutense de Madrid, 28040 Madrid, Spain
e-mail: parrondo@ucm.es

J. Tabanera-Bravo
e-mail: jorgetab@ucm.es

F. Fedele · N. Ares
Department of Engineering Science, University of Oxford, OX1 3PJ, Oxford, UK
e-mail: federico.fedele@eng.ox.ac.uk

N. Ares
e-mail: natalia.ares@eng.ox.ac.uk

X. Bouju and C. Joachim (eds.), *Crossroad of Maxwell Demon*, Advances in Atom and Single Molecule Machines, https://doi.org/10.1007/978-3-031-57904-2_1

molecules from the left gas to pass to the right gas and slow molecules from the right gas to pass to the left gas. By doing so, the demon establishes a heat flow from the left to the right and can sustain this flow even against a temperature gradient, without the need of performing work [1].

The crucial aspect of the Maxwell demon is the information gathered during the measurement and subsequently used in the operation of the gate. In this sense, the demon is implementing a feedback control protocol. The first problem that the Maxwell demon poses to thermodynamics is how to incorporate information into the second law or, in other words, how to find a second law for feedback processes [2–4]. The solution was partly given by a stylized version of the demon due to Szilárd [5], where an external agent performs a binary measurement, gathering one bit of information, and extracts a work $kT \ln 2$ from a thermal bath at temperature T, k being the Boltzmann constant. A more general relationship between the information obtained by the external agent in the measurement and the extracted work (or, equivalently, the entropy reduction) was recently revealed by Sagawa and Ueda [6], using an interesting concept from information theory: the mutual information between two random variables [7], which we explain in detail in Sect. 2.

The second problem posed by the Maxwell demon is the restoration of the second law by considering the physical nature of the demon [3] and the heat dissipation or entropy cost associated with its operation. The common wisdom was that there is an unavoidable entropy cost in the measurement [1] but, in the 1970s, Charles Bennett applied Rolf Landauer's ideas on the thermodynamics of information devices [8] and introduced an unexpected twist by showing that, in certain setups, the measurement can be carried out with no heat dissipation [9]. In these cases, the restoration of the second law is due to an unavoidable heat dissipation in the resetting of the demon's memory, i.e., in the erasure of the information registered in the measurement. These ideas formed the basis for a new field called *thermodynamics of information*, which has benefited from more recent results in non-equilibrium statistical mechanics and stochastic thermodynamics [2–4].

The purpose of this chapter is not to provide a review of the thermodynamics of information. The reader can find a more exhaustive introduction to the field in Refs. [2–4]. Here we focus on the results of the theory that can be applied to autonomous machines. We first introduce in Sect. 2 the main concepts from information theory, namely, Shannon entropy and the mutual information between two random variables. In Sect. 3, we discuss the relationship between Shannon entropy and thermodynamic entropy in stochastic systems described by master equations. There we show that the Shannon entropy of a non-equilibrium system in contact with thermal and particle reservoirs possesses similar properties as the thermodynamic entropy and, in particular, can be used to define a non-equilibrium free energy that assesses the amount of energy that can be converted into useful work. These tools are used in Sect. 4 to analyze the thermodynamic properties of different setups where information plays a relevant role: feedback processes and autonomous and driven bipartite systems. Finally, in Sect. 5 we apply these results to a specific setup based on carbon nanotubes where mechanical oscillations and electron transport are coupled.

2 Shannon Entropy and Mutual Information

In 1948, Claude E. Shannon introduced a measure of the uncertainty of a random variable X, which could describe, for instance, a message in a communication channel, the microstate of a physical system, or the outcome of a measurement [7]. The Shannon uncertainty, also known as Shannon entropy, of a random variable X is defined as

$$S(X) = S(\rho_X) \equiv -\sum_x \rho_X(x) \log \rho_X(x) \quad (1)$$

where $\rho_X(x)$ is the probability distribution of X and the sum runs over all possible values x that the variable can take on (for continuous random variables, one must replace the sum with an integral, and the entropy is defined up to an additive constant [7]). We have used two alternative notations for the Shannon entropy or uncertainty. The first one, $S(X)$, is commonly used in information theory and more appropriate for situations with several random variables (see below). The second one, $S(\rho_X)$ expresses explicitly the dependence of Shannon entropy on the probability distribution ρ_X and is more usual in statistical physics. Shannon uncertainty is measured in *bits* if the logarithm in Eq. (1) is in base 2, and in *nats* if the natural logarithm is used. Moreover, if we multiply (1) by the Boltzmann constant k, Shannon entropy is then measured in units of physical entropy (energy divided by absolute temperature). In this work, we will always express Shannon entropy or Shannon uncertainty in these units or, more precisely, in nats times the Boltzmann constant k.

Information is defined as the reduction in the uncertainty of a random variable X when we inquire about it by, for instance, performing a measurement or asking a question to somebody who knows X. The outcome of the measurement or the answer to the question is a random variable Y, not necessarily equal to X. For example, if somebody knows the result $X = 1, 2, \ldots, 6$ of a roll of a die and we ask her whether X is odd, then the answer Y takes on only two possible values, 'yes' or 'no'. The information that the answer Y provides about the random variable X is the reduction of its uncertainty, i.e.,

$$I(X;Y) = S(X) - S(X|Y), \quad (2)$$

and is called the *mutual information* between X and Y. Here $S(X|Y)$ is the uncertainty of the posterior distribution $\rho_{X|Y}(x|y)$ averaged over the measurement outcomes y:

$$S(X|Y) = \sum_y \rho_Y(y) \left[-\sum_x \rho_{X|Y}(x|y) \log \rho_{X|Y}(x|y) \right]. \quad (3)$$

Using the relation $\rho_{XY}(x, y) = \rho_{X|Y}(x|y)\rho_Y(y)$ one obtains

$$I(X;Y) = \sum_{x,y} \rho_{XY}(x, y) \log \frac{\rho_{XY}(x, y)}{\rho_X(x)\rho_Y(y)}. \quad (4)$$

We observe that mutual information is symmetric, that is, the information that X provides about Y equals the information that Y provides about X. In addition, mutual information is a measure of the correlation between the two variables: $I(X; Y) \geq 0$ and vanishes when X and Y are uncorrelated. Another way of writing mutual information is

$$I(X; Y) = S(X) + S(Y) - S(X, Y) \geq 0 \tag{5}$$

which shows that entropy is sub-additive: $S(X, Y) = S(X) + S(Y) - I(X; Y) \leq S(X) + S(Y)$. The equality is met if and only if the two random variables X and Y are statistically independent.

If Y is a deterministic function of X, which is the case of an error-free measurement, then $S(Y|X) = 0$ and $I(X; Y) = S(Y) - S(Y|X) = S(Y)$, i.e., the mutual information is the Shannon entropy of the measurement outcome.

Equations (2) and (5) express, respectively, two properties of mutual information that are relevant in thermodynamics. The first property (2) states that mutual information is equal to the decrease in entropy resulting from a measurement. The second property (5) is useful for connecting the entropy $S(X, Y)$ of a bipartite system described by the random variables (X, Y) with the entropies of its subsystems, $S(X)$ and $S(Y)$. However, in order to understand the physical implications of measurements and other processes involving information, it is still necessary to clarify the relationship between Shannon entropy and thermodynamic entropy.

3 Shannon and Thermodynamic Entropies

Shannon entropy coincides with thermodynamic entropy for equilibrium states. Nevertheless, the identification of both magnitudes for systems far from equilibrium is not always correct. For instance, the Shannon entropy of a probability distribution over the microstates of a system is invariant under Hamiltonian evolution and, consequently, does not reproduce the increase in thermodynamic entropy dictated by the second law. However, in the case of non-equilibrium systems in contact with thermal baths and particle reservoirs, Shannon entropy has the same properties as thermodynamic entropy. This can be proved in the context of Hamiltonian evolution [4], as well as stochastic evolution given by Langevin and/or master equations [10, 11]. We now discuss this important result for a discrete system ruled by a master equation, following Ref. [10], and later on we extend the discussion to a bipartite hybrid system (continuous and discrete) describing the coupling between the electron transport through a carbon nanotube and its mechanical oscillations.

3.1 Entropy Production in Discrete Systems

Consider a system with discrete states i in contact with several thermostats and chemostats (particle reservoirs) at different temperatures and chemical potentials. The stochastic evolution of the system is described by a master equation:

$$\dot{\rho}(i,t) = \sum_j \left[\rho(j,t)\Gamma_{j\to i} - \rho(i,t)\Gamma_{i\to j}\right] \tag{6}$$

where $\rho(i,t)$ is the probability that the system is in state i at time t and $\Gamma_{i\to j}$ is the probability transition rate from state i to state j (more precisely, the probability to observe the transition $i \to j$ in a short time interval of duration Δt is $\rho(i,t)\Gamma_{i\to j}\Delta t$). The master equation can also be written as

$$\dot{\rho}(i,t) = \sum_j J_{j\to i}(t) \tag{7}$$

where

$$J_{j\to i}(t) \equiv \rho(j,t)\Gamma_{j\to i} - \rho(i,t)\Gamma_{i\to j} \tag{8}$$

is the net probability current from state j to state i. The change in the Shannon entropy of the system can be written as

$$\begin{aligned}\dot{S}(t) &= -\frac{d}{dt}k\sum_i \rho(i,t)\ln\rho(i,t) = -k\sum_i \frac{d\rho(i,t)}{dt}\ln\rho(i,t)\\ &= -k\sum_{i,j} J_{j\to i}(t)\ln\rho(i,t) = k\sum_{i>j} J_{i\to j}(t)\ln\frac{\rho(i,t)}{\rho(j,t)}\end{aligned} \tag{9}$$

where we have used the normalization of the probability distribution: $\sum_i d\rho(i,t)/dt = 0$ in the first step and the anti-symmetry of the current: $J_{i\to j}(t) = -J_{j\to i}(t)$ in the last step. The expression indicates that each transition, $i \to j$, contributes to the increase in Shannon entropy as $k\ln[\rho(i,t)/\rho(j,t)]$.

To relate Shannon entropy with thermodynamic entropy, we need to specify the physical nature of the system and its environment. For simplicity, we will assume that the states of the system do not have internal entropy and their free energy equals their energy E_i. The system exchanges energy with several thermal baths at different temperatures and some transitions can be mediated by particle reservoirs with different chemical potentials. In this case, the transition rates obey local detailed balance[1]:

[1] In some setups, a transition can follow different pathways involving different reservoirs. In those situations, local detailed balance is fulfilled by the rates of each pathway [12]. This is the case of a quantum dot in contact with two reservoirs, as we discuss in Sect. 5.1.

$$\frac{\Gamma_{i\to j}}{\Gamma_{j\to i}} = e^{-\beta_{ji}\left(E_j - E_i + \Delta F_{i\to j}^{(\text{env})}\right)} \tag{10}$$

where $\beta_{ji} = \beta_{ij} = 1/(kT_{ij})$ is the inverse temperature of the bath that induces the transitions between i and j and $\Delta F_{i\to j}^{(\text{env})}$ is the change in the free energy of the environment due to the transition from i to j. Since $F = E - TS$ in each reservoir and $E_i - E_j = \Delta E_{i\to j}^{(\text{env})}$ is the energy dissipated to the reservoir at temperature T_{ij}, the total change in the thermodynamic entropy of the environment due to a transition from i to j is given by

$$\Delta S_{i\to j}^{(\text{env})} = \frac{E_i - E_j - \Delta F_{i\to j}^{(\text{env})}}{T_{ij}} = k \ln \frac{\Gamma_{i\to j}}{\Gamma_{j\to i}}. \tag{11}$$

Then, the entropy of the environment changes at a rate

$$\dot{S}_{\text{env}}(t) = \sum_{i>j} J_{i\to j}(t) \Delta S_{i\to j}^{(\text{env})} = k \sum_{i>j} J_{i\to j}(t) \ln \frac{\Gamma_{i\to j}}{\Gamma_{j\to i}}. \tag{12}$$

If we sum up the change in the thermodynamic entropy of the environment (12) and in the Shannon entropy of the system (9), we get a total entropy production that reads

$$\dot{S}_{\text{prod}}(t) \equiv \dot{S}_{\text{env}}(t) + \dot{S}(t) = k \sum_{i>j} J_{i\to j}(t) \ln \frac{\rho(i,t)\Gamma_{i\to j}}{\rho(j,t)\Gamma_{j\to i}} \geq 0. \tag{13}$$

Each term in the sum is positive because the logarithm and the current have the same sign. Hence, $\dot{S}_{\text{prod}}(t) \geq 0$, and the equality is reached only in the case of global detailed balance, where all currents vanish. In summary, we have proved that the Shannon entropy can be added to the environment entropy to yield a positive entropy production. Notice that the environment entropy is well-defined if the environment consists of equilibrium reservoirs. In this case, the change in entropy ΔS is given by the usual expressions in equilibrium thermodynamics: $\Delta S = \Delta E/T + \mu \Delta N/T$, where ΔE and ΔN are the changes in the energy and number of particles of the reservoir, respectively, T is its temperature, and μ its chemical potential. Hence, for systems in contact with equilibrium reservoirs, Shannon entropy possesses the same properties as thermodynamic entropy even for non-equilibrium states and processes.

3.2 *Energetics and Non-equilibrium Free Energy*

All the previous equations hold for time-dependent transition rates $\Gamma_{i\to j}(t)$. In particular, they are valid for driven physical systems, with energies $E_i(\lambda)$ depending on a parameter λ that an external agent modifies according to a given protocol $\lambda(t)$ with

$t \in [0, \tau]$, τ being the duration of the protocol. In this case, we can split the change in the average energy of the system $E(t) \equiv \sum_i p(i,t)E_i(\lambda(t))$ into two terms:

$$\dot{E}(t) = \sum_i \left[\frac{dp(i,t)}{dt} E_i(\lambda(t)) + p(i,t) \frac{dE_i(\lambda(t))}{dt} \right]. \tag{14}$$

The first term is the exchange of energy with the environment. It can be expressed as a sum of the energy transfer in each transition, following an analogous argument as the one used in Eq. (9):

$$\sum_i \frac{dp(i,t)}{dt} E_i(\lambda(t)) = \sum_{i>j} J_{i\to j}(t) \left[E_j(\lambda(t)) - E_i(\lambda(t)) \right]. \tag{15}$$

This quantity can be further split into two terms using (11):

$$\begin{aligned} \sum_i \frac{dp(i,t)}{dt} E_i(\lambda(t)) &= -\sum_{i>j} J_{i\to j}(t) \left[T_{ij} \Delta S^{(\text{env})}_{i\to j} + \Delta F^{(\text{env})}_{i\to j} \right] \\ &= \dot{Q}(t) + \dot{W}_{\text{chem}}(t). \end{aligned} \tag{16}$$

Here, the heat $\dot{Q}(t) \equiv -\sum_{i>j} J_{i\to j}(t) T_{ij} \Delta S^{(\text{env})}_{i\to j}$ is the energy that changes the entropy of the environment, whereas the rest of the energy is usually called chemical work, $\dot{W}_{\text{chem}}(t) \equiv -\sum_{i>j} J_{i\to j}(t) \Delta F^{(\text{env})}_{i\to j}$. This work is energy that the reservoirs transfer to the system without affecting their entropy. Chemical work is relevant when some transitions are mediated by chemostats, as occurs in chemical motors [13], as well as when the system and the reservoirs exchange particles. This is the case of the setup that we discuss in Sect. 5, where a quantum dot exchanges electrons with two electrodes.

Finally, the last term in (14) is the energy that the system takes from the external agent, and is usually called mechanical work or simply work:

$$\dot{W}(t) \equiv \sum_i p(i,t) \frac{dE_i(\lambda(t))}{dt}. \tag{17}$$

These energy exchanges can be combined into a single equation, which is the mathematical expression of the first law of thermodynamics[2]:

$$\dot{E}(t) = \dot{Q}(t) + \dot{W}(t) + \dot{W}_{\text{chem}}(t). \tag{18}$$

For an isothermal process (i.e., when all the baths are at the same temperature T), the heat is $\dot{Q}(t) = -T\dot{S}_{\text{env}}(t)$. In this case, from the first law (18) and the second law (13), one can immediately obtain

[2] Note that the overdot on heat and work does not indicate the time derivative of a given quantity, but a rate of energy exchange.

$$T\dot{S}_{\text{prod}}(t) = -\dot{Q}(t) + T\dot{S}(t) = \dot{W}(t) + \dot{W}_{\text{chem}}(t) - \dot{\mathcal{F}}(t) \geq 0 \tag{19}$$

where

$$\mathcal{F}(t) \equiv E(t) - TS(t) \tag{20}$$

is called *non-equilibrium free energy*. It can be defined in general for any physical system with Hamiltonian $\mathcal{H}(x)$ and probabilistic state $\rho_X(x)$ as the average energy minus the Shannon entropy multiplied by the temperature [3]:

$$\mathcal{F}(\rho_X, \mathcal{H}) \equiv \langle \mathcal{H}(X) \rangle_{\rho_X} - TS(\rho_X). \tag{21}$$

We will also use the notation $\mathcal{F}(X)$ for the non-equilibrium free energy of a system whose microstate is the random variable X with probability distribution $\rho_X(x)$. This definition is valid for probability distributions $\rho(i)$ defined over micro- or meso-states i with energy E_i, as well as for quantum density matrices. Like in the case of isothermal processes connecting equilibrium states, the non-equilibrium free energy tells us the amount of work that can be extracted from non-equilibrium states. Indeed, integrating Eq. (19) over the whole process $t \in [0, \tau]$, we obtain

$$W_{\text{tot}} \geq \Delta\mathcal{F} \tag{22}$$

where $W_{\text{tot}} = W + W_{\text{chem}}$ is the total work performed by the external agent and the reservoirs on the system, and

$$\Delta\mathcal{F} = \mathcal{F}[\rho_X(x;\tau), \mathcal{H}(x;\tau)] - \mathcal{F}[\rho(x;0), \mathcal{H}(x;0)] \tag{23}$$

is the increase in non-equilibrium free energy. Here $\rho_X(x;t)$ and $\mathcal{H}(x;t)$ are, respectively, the probabilistic state of the system and the Hamiltonian at time t. Thus, we can extract work, $W_{\text{tot}} < 0$, by driving a system from a non-equilibrium state with high free energy to a state with lower non-equilibrium free energy. See [3] for a specific protocol that is able to extract the maximum amount of work $-\Delta\mathcal{F}$. These results are also valid for continuous systems ruled by Langevin-like equations, as well as for a wide family of Hamiltonian systems [4].

4 Thermodynamics of Information

We have proven that the change in Shannon entropy of a system can be added to the change in thermodynamic entropy of its environment, resulting in a total entropy production that is positive and only vanishes for reversible processes, as shown by Eq. (13). Shannon entropy is also used to define non-equilibrium free energy in Eq. (21), which establishes a lower bound on the work required to complete an isothermal process connecting non-equilibrium states, as shown by Eq. (22).

In this section, we analyze the interplay between information, entropy, and energy in different physical situations, utilizing the properties of the Shannon entropy introduced in Sect. 2. Processes in which information plays a significant role always involve non-equilibrium states, such as those resulting from a measurement. In this sense, thermodynamics of information can be regarded as a branch of non-equilibrium thermodynamics that deals with a specific class of non-equilibrium states.

We begin by discussing the effect of feedback on the second law of thermodynamics, which is present in the original Maxwell demon and the Szilárd engine. Next, we incorporate the physical nature of the demon to restore the original second law. For this purpose, we study bipartite systems, constituted by the machine and the demon. We first consider bipartite systems driven by an external agent who switches on and off the interactions between the machine and the demon and restores the state of the demon. This external agent drives the system according to a non-feedback protocol, that is, they cannot measure the bipartite system nor use any information of its actual state. In this way, the bipartite system undergoes a normal non-feedback thermodynamic process and the second law is restored. Nevertheless, the process can be interpreted as the demon gathering information from the machine and operating according to that information. Finally, we consider the case of bipartite autonomous systems in a non-equilibrium steady state (NESS). In this case, which is probably the most relevant for nanomachines, we can still interpret one of the two systems as gathering information from the other and implementing some kind of feedback, using a new concept: information flows.

4.1 Isothermal Feedback Processes

The simplest setup in which information affects the limitations imposed by thermodynamics is a process where an external agent performs a measurement on a system and implements a protocol based on the measurement outcome Y. According to (2), the measurement reduces the Shannon entropy of the system by an amount $I(X;Y)$, where X is the microstate of the system. If the measurement does not affect the energy of the system, this reduction in entropy leads to an increase in the non-equilibrium free energy:

$$\Delta\mathcal{F}_{\text{meas}} \equiv \mathcal{F}_{\text{post}} - \mathcal{F}_{\text{pre}} = TI(X;Y). \tag{24}$$

where $\mathcal{F}_{\text{post}}$ and $\mathcal{F}_{\text{pre}}$ are, respectively, the non-equilibrium free energies immediately before and after the measurement.

In principle, this increase may require work and heat dissipation, as we will discuss in the next subsection. Here, we are solely interested in finding how the second law (22) should be modified to incorporate the information acquired in the measurement. The modification simply consists in adding the increment in free energy $\Delta\mathcal{F}_{\text{meas}}$ to the available free energy that can be converted into work. More precisely, the work before the measurement obeys $W_{\text{pre}} \geq \mathcal{F}_{\text{pre}} - \mathcal{F}_0$, where $\mathcal{F}_0$ is the initial free energy, and the work after the measurement is bound as $W_{\rm post} \geq \mathcal{F}_\tau - \mathcal{F}_{\rm post}$, where $\mathcal{F}_\tau$ is the free energy at the end of the process. If we do not take into account the work performed in the measurement, the non-equilibrium second law (22), that is, the total work necessary to complete the process is bound as [4]

$$W_{\rm fb} = W_{\rm pre} + W_{\rm post} \geq \Delta\mathcal{F} - TI(X;Y). \tag{25}$$

For a cyclic process, $\Delta\mathcal{F} = 0$ but, if the measurement provides some information about the microstate of the system, one can extract work ($W_{\rm fb} < 0$), in apparent contradiction with the second law of thermodynamics. A simple example is the aforementioned Szilárd engine [5], in which a particle is in a container of volume V whose walls are at temperature T. A piston is inserted in the middle of the container and the demon measures in which half the particle lies, i.e., the outcome Y is either left or right with equal probability. If the measurement has no errors, then $I(X;Y) = S(Y) = 1$ bit $= k \ln 2$. This information can be used to implement a reversible isothermal expansion extracting an energy $kT \ln 2$ from the thermal bath, in agreement with the second law (25). The process is sketched in Fig. 1.

An interesting problem is whether the bound in (25) can be reached. This is the case of the Szilárd engine, for instance. Protocols that saturate the bound are called optimal feedback protocols or optimal Maxwell demons [14]. In [15], we show that they must be operationally reversible, i.e., the system must go through the same probabilistic states when the protocol is run forward and backward in time. This implies that optimal protocols are quasistatic, as in standard thermodynamics, but must also reproduce post-measurement probabilistic states, a requirement that can be difficult to fulfill and yields highly non-trivial protocols [14].

4.2 Driven Bipartite Systems

We now consider the physical nature of the demon. As a physical system, the demon is described by a microscopic state z comprising the positions and momenta of all its molecules. Its evolution and interaction with the system are ruled by a Hamiltonian $\mathcal{H}_{\rm demon}(z) + \mathcal{H}_{\rm int}(x, z; t)$, where x denotes the microstate of the controlled system.

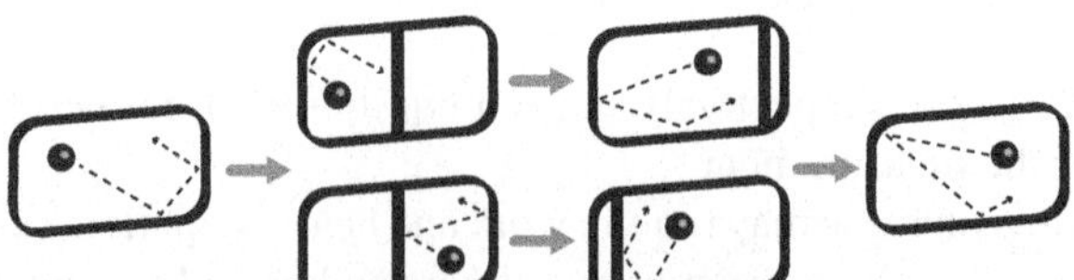

Fig. 1 Sketch of the Szilárd engine: a piston is inserted in the middle of a box containing a single particle gas. By measuring the half where the particle lies after the insertion, one can perform a reversible isothermal expansion and extract an energy $kT \ln 2$ from the thermal bath

Still, an external agent is needed to switch on and off the interaction between the demon and the system in the measurement and in the feedback operation. This external agent, however, carries out a protocol on the bipartite system (system+demon) that does not use any information about the state of the system. Hence, the usual second law (22) holds. According to Eq. (5), the entropy of the bipartite system can be written as

$$S(X(t), Z(t)) = S(X(t)) + S(Z(t)) - I(X(t); Z(t)), \tag{26}$$

and the non-equilibrium free energy is

$$\mathcal{F}(X(t), Z(t)) = \mathcal{F}(X(t)) + \mathcal{F}(Z(t)) + \langle \mathcal{H}_{\text{int}}(X, Z; t)\rangle + T I(X(t); Z(t)). \tag{27}$$

Consider now the feedback isothermal process depicted in Fig. 2. We assume that the demon and the system start in random states X and Z, respectively, without any interaction $\mathcal{H}_{\text{int}} = 0$ and correlation, $I(X; Z) = 0$. The measurement is an interaction between the demon and the system that, ideally, does not alter the state X of the system. The state of the demon does change to Z' and correlates with the state of the system generating a mutual information $I(X; Z') > 0$. The outcome of the measurement Y is a function of the microstate of the demon Z': $Y = y(Z')$. If all the information about X is provided by the measurement outcome Y, then the conditional probabilities obey $\rho_{X|Z'}(x|z') = \rho_{X|Y}(x|y(z'))$, and it is not hard to prove that the mutual information verifies

$$I(X; Z') = I(X; Y). \tag{28}$$

If the interaction is switched off after the measurement, then the change in non-equilibrium free energy due to the measurement is $\Delta\mathcal{F}_{\text{meas}} = \mathcal{F}(Z') + T I(X; Z') - \mathcal{F}(Z)$. This change requires a work bound by $\Delta\mathcal{F}_{\text{meas}}$:

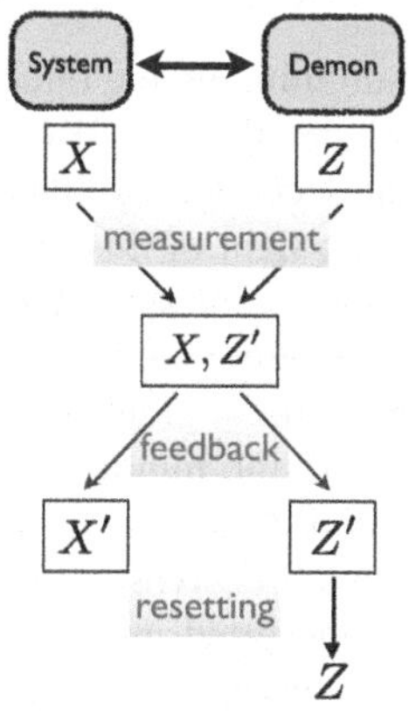

Fig. 2 Evolution of the state of the system and the demon in the different stages of a feedback process

$$W_{\text{meas}} \geq \Delta\mathcal{F}_{\text{meas}} = \Delta\mathcal{F}_{\text{demon}} + TI(X;Y) \tag{29}$$

where $\Delta\mathcal{F}_{\text{demon}} = \mathcal{F}(Z') - \mathcal{F}(Z)$ is the change in the non-equilibrium free energy of the demon due to the measurement. Notice that the measurement can be carried out without any work and heat dissipation if the free energy of the demon decreases and compensates for the term due to the mutual information [9].

After the measurement, the demon implements the feedback process, where the system is driven from state X to state X'. We assume that the demon state does not change during the feedback protocol. If all correlations are exhausted, meaning that $I(X';Z') = 0$, and the interaction is switched off at the end of the process, then the change in non-equilibrium free energy in the feedback process is $\Delta\mathcal{F}_{\text{fb}} = \Delta\mathcal{F}_{\text{system}} - TI(X;Z')$, where $\Delta\mathcal{F}_{\text{system}} = \mathcal{F}(X') - \mathcal{F}(X)$, and the work is bound as:

$$W_{\text{fb}} \geq \Delta\mathcal{F}_{\text{fb}} = \Delta\mathcal{F}_{\text{system}} - TI(X;Y). \tag{30}$$

This inequality coincides with (25), as expected. However, we have obtained a new result: to achieve the equality in both equations, (25) and (30), it is necessary to exploit all the correlations $I(X;Z') = I(X;Y)$ generated in the measurement. Notice also that, in the derivation of (25), mutual information appeared as the reduction of uncertainty in the system due to the measurement, whereas here it is interpreted as a measure of the correlation between the demon and the system.

For a cycle, $\Delta\mathcal{F}_{\text{system}} = 0$, and the amount of work that can be extracted from the thermal bath is $TI(X;Y)$. If we perform a measurement without work, i.e., decreasing the free energy of the demon by $\Delta\mathcal{F}_{\text{demon}} = -TI(X;Y)$, then we have an information engine that can extract work from a single thermal bath. However, there is no contradiction with the second law, since the final state of the demon Z' is not equal to the initial state Z. In other words, the process is not cyclic unless we add a third step where the memory of the demon is reset. The resetting changes the free energy as $-\Delta\mathcal{F}_{\text{demon}}$ and needs a work that verifies

$$W_{\text{reset}} \geq -\Delta\mathcal{F}_{\text{demon}}. \tag{31}$$

Summing up the three bounds, (29), (30), and (31), we conclude that the total work $W_{\text{tot}} = W_{\text{meas}} + W_{\text{fb}} + W_{\text{reset}}$ is always positive in a cycle ($X' = X$), as dictated by the second law of thermodynamics.

In light of this discussion, we conclude that the Maxwell demon consists of the creation and subsequent exploitation of correlations between two systems: one acting as a demon and the other as a "working substance". This interpretation applies to non-autonomous nanomachines described by bipartite systems: the intervals of time where the mutual information between the two systems, $I(X(t);Y(t))$, increases can be identified as measurements, whereas those intervals where the mutual information decreases can be interpreted as feedback processes where the information is converted into work [16]. This analysis can reveal the efficiency of each stage, measurement, feedback, and resetting, as well as whether the correlations generated by the measurement are optimally exploited in the feedback stage of the process.

Another interesting aspect of the previous analysis is that we can leave out the last stage of the cycle —resetting the demon's memory—if the memory has initially a high non-equilibrium free energy. In this case, we can run the process many times extracting work and degrading the memory. In other words, an ordered memory can be used as fuel. These information reservoirs have been studied in general by Barato and Seifert in [17]. Mandal and Jarzynski devised a specific kinetic model that can extract work from a single thermal bath using as fuel an ordered memory on a tape [18].

If we want to extend the previous analysis to bipartite systems in contact with several reservoirs at different temperatures, we must use entropy instead of free energy. The combination of the second law (13) and the expression for the Shannon entropy of a bipartite system (X, Y) in terms of the mutual information (26) yields

$$\dot{S}_{\text{prod}} = \dot{S}(X) + \dot{S}(Y) - \dot{I}(X; Y) + \dot{S}_{\text{env}} \geq 0, \tag{32}$$

where now the change in the entropy of the environment depends on the physical properties of the reservoirs.

4.3 *Autonomous Bipartite Systems: Information Flow*

Consider an autonomous bipartite system (X, Y) in a non-equilibrium steady state (NESS). Eqs (26), (27), and (32) are still valid but useless since all the terms are constant in the first two equations and vanish in the latter. Information flows try to overcome this problem by focusing on the changes in mutual information due to the evolution of each part of the bipartite system. They are defined as [16, 19–21]

$$\dot{I}^X(t) \equiv \left.\frac{d}{dt'}\right|_{t'=t} I(X(t'); Y(t)); \quad \dot{I}^Y(t) \equiv \left.\frac{d}{dt'}\right|_{t'=t} I(X(t); Y(t')). \tag{33}$$

Until now, we have used the term 'bipartite system' to refer to any system with two parts described by microstates (X, Z) or (X, Y).[3] To establish a connection between the information flows (33) and the change in mutual information, we need to impose further restrictions on this definition. In a bipartite discrete system, transitions where the state of the two subsystems changes simultaneously are not allowed, i.e.,

$$\Gamma_{(x,y)\to(x',y')} = 0 \quad \text{if } x \neq x' \text{ and } y \neq y'. \tag{34}$$

Hence, one can consider separately the transitions where each of the subsystems changes:

$$\Gamma^y_{x\to x'} \equiv \Gamma_{(x,y)\to(x',y)} \tag{35}$$

[3] In this subsection, and in the rest of the chapter, we no longer distinguish between the state of the demon and the measurement outcome.

and

$$\Gamma^{x}_{y\to y'} \equiv \Gamma_{(x,y)\to(x,y')}. \tag{36}$$

The master equation (7) can be split into two sums corresponding to the evolution of systems X and Y, respectively,

$$\dot{\rho}(x, y; t) = \sum_{x', y'} \left[J^{y}_{x'\to x}(t) + J^{x}_{y'\to y}(t) \right] \tag{37}$$

where the current due to changes in X is

$$J^{y}_{x'\to x}(t) \equiv \Gamma^{y}_{x'\to x}\rho(x', y; t) - \Gamma^{y}_{x\to x'}\rho(x, y; t) \tag{38}$$

and the current corresponding to system Y reads

$$J^{x}_{y'\to y}(t) \equiv \Gamma^{x}_{y'\to y}\rho(x, y'; t) - \Gamma^{x}_{y\to y'}\rho(x, y; t). \tag{39}$$

Now every term in Eq. (32) can be decomposed into two parts accounting for the contributions of transitions in systems X and Y, respectively. For transitions in system X, this contribution is

$$\dot{S}^{X}_{\text{prod}}(t) \equiv \dot{S}(X(t)) - \dot{I}^{X}(t) + \dot{S}^{X}_{\text{env}}(t). \tag{40}$$

Each derivative here is expressed in terms of $J^{y}_{x'\to x}(t)$. Using similar arguments as in the derivation of Eq. (9), we express the change in Shannon entropy as

$$\dot{S}(X(t)) = k \sum_{y} \sum_{x>x'} J^{y}_{x'\to x}(t) \ln \frac{\rho(x', t)}{\rho(x, t)}. \tag{41}$$

It is not difficult to prove that the information flow defined in (33) can be written as

$$\begin{aligned} \dot{I}^{X}(t) &= k \sum_{x,y} \sum_{x'} J^{y}_{x'\to x}(t) \ln \frac{\rho(x, y; t)}{\rho(x, t)\rho(y, t)} \\ &= k \sum_{y} \sum_{x>x'} J^{y}_{x'\to x}(t) \ln \frac{\rho(x, y; t)\rho(x', t)}{\rho(x', y; t)\rho(x, t)}, \end{aligned} \tag{42}$$

where we have used the anti-symmetry of the current $J^{y}_{x'\to x} = -J^{y}_{x\to x'}$. Finally, the entropy increase in the environment due to transitions in X is given by

$$\dot{S}^{X}_{\text{env}}(t) = k \sum_{y} \sum_{x>x'} J^{y}_{x'\to x}(t) \ln \frac{\Gamma^{y}_{x'\to x}}{\Gamma^{y}_{x\to x'}}. \tag{43}$$

This expression is a consequence of local detailed balance, as in (12). Introducing Eqs. (41), (42), and (43) into (40), we obtain

$$\dot{S}^X_{\text{prod}}(t) = k \sum_y \sum_{x>x'} J^y_{x'\to x}(t) \ln \frac{\rho(x', y; t)\Gamma^y_{x'\to x}}{\rho(x, y; t)\Gamma^y_{x\to x'}} \geq 0. \tag{44}$$

As in Eq. (13), the inequality holds because every term in the sum is positive, given that the current and the logarithm have the same sign.

In the stationary state, the marginal Shannon entropies are constant: $\dot{S}(X) = \dot{S}(Y) = 0$, as well as the mutual information: $\dot{I}(X; Y) = \dot{I}^X + \dot{I}^Y = 0$. Hence, there is a single quantity characterizing the information flow: $\dot{I}^Y = -\dot{I}^X$. The sign of the information flow indicates the role played by each subsystem. In the previous section, we showed that mutual information increases during the measurement and decreases during feedback. Additionally, we observed that, in an ideal measurement, the state of the demon changes and becomes correlated with the state of the system, without altering the latter. Conversely, in the feedback process, the system changes while, ideally, the state of the demon remains constant. We can apply these observations to the autonomous bipartite system. For instance, if $\dot{I}^Y$ is positive and $\dot{I}^X$ is negative, then the evolution of system Y, keeping X constant, leads to an increase in mutual information, while the evolution of system X decreases mutual information. By comparing this scenario with the driven case, we can identify system Y as the demon and system X as the one on which the demon measures and operates.

The inequality (44) together with the definition of the entropy production (40) due to the evolution of X, and the analogous equation for system Y, yield the following local second laws for the bipartite system in a NESS [16, 20]:

$$\begin{aligned} \dot{S}^X_{\text{prod}} &= \dot{S}^X_{\text{env}} + \dot{I}^Y \geq 0 \\ \dot{S}^Y_{\text{prod}} &= \dot{S}^Y_{\text{env}} - \dot{I}^Y \geq 0. \end{aligned} \tag{45}$$

This is the main result of this section. The sum of the two inequalities yields the usual second law for a NESS, namely, that the entropy in the environment increases, $\dot{S}_{\text{env}} \geq 0$. The two inequalities in (45) impose an extra limitation on the functioning of the system and reveal how the bipartite system operates as an information machine [16, 20]. If $\dot{I}^Y$ is positive, Y acts as a demon that measures on X. For this purpose, Y uses the entropic resources of its environment, $\dot{S}^Y_{\text{env}} \geq \dot{I}^Y \geq 0$. As a consequence, system X can reduce the entropy of its reservoirs: $\dot{S}^X_{\text{env}} \geq -\dot{I}^Y$. Therefore, subsystem X could apparently defy the second law of thermodynamics due to the information flow from subsystem Y.

This use of thermodynamic resources can be represented by the flow diagram depicted in Fig. 3, analogous to the so-called Sankey diagrams that represent the transfers and leaks of a given resource in engines and other processes and devices. In our case, the increase of entropy in the Y-environment is transferred to system X through the information flow $\dot{I}^Y$ and used to decrease the entropy of the X-environment. The entropy production in each subsystem, $\dot{S}^X_{\text{prod}}$ and $\dot{S}^Y_{\text{prod}}$, can be

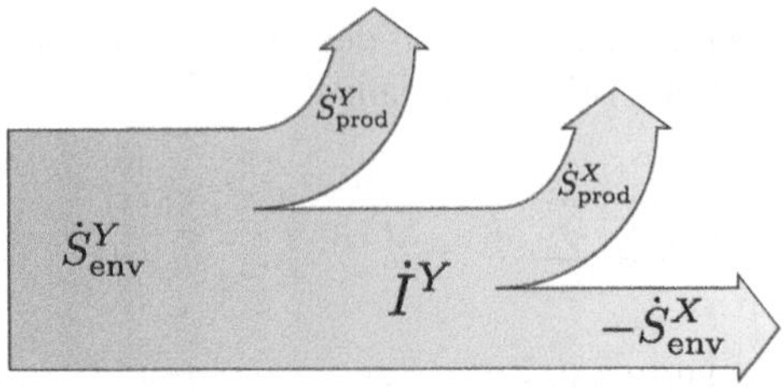

Fig. 3 The entropic Sankey diagram of a bipartite system. The increase of entropy in the Y-environment, $\dot{S}^Y_{\rm env} > 0$, is used to decrease the entropy in the X-environment, $\dot{S}^X_{\rm env} < 0$. The transfer of this capacity to decrease entropy from system Y to system X occurs through the information flow $\dot{I}^Y > 0$, while the entropy production in each system can be interpreted as leaks in this transfer [16]. All the quantities in the diagram are positive and represent the transfer of a conserved quantity. For instance, $\dot{S}^Y_{\rm env} = \dot{S}^Y_{\rm prod} + \dot{I}^Y$. Hence, the diagram is a graphic representation of the local second laws (45) for the specific case $\dot{S}^Y_{\rm env} > 0$, $\dot{I}^Y > 0$ and $\dot{S}^X_{\rm env} < 0$

considered leaks in this transfer of entropy. However, this entropic Sankey diagram differs from the more usual one for engines, which describes transfers and losses of energy. In the entropic diagram of Fig. 3, the resource is the increase of entropy in one of the reservoirs, $\dot{S}^Y_{\rm env} > 0$. In a chemical motor, for instance, the reservoir could be a chemostat that indeed provides free energy to the motor; however, in a thermal motor, the reservoir where the entropy increases is the cold bath, where heat is dissipated. Moreover, the output in the diagram, $-\dot{S}^X_{\rm env}$ does not always coincide with or is proportional to the functional purpose of the machine, namely, the work in an engine or the extracted heat in a heat pump. An extreme case is the heat flow between two thermal baths in contact. Clearly, this is not a useful thermal machine; however, one could draw an entropic Sankey diagram where the resource is the entropy increase in the cold bath and the output is the entropy decrease in the hot bath. Despite this, the diagram is useful to analyze the performance of certain devices, as we show in the next section.

For isothermal chemical machines, the local second laws (45) can be combined with the conservation of energy in the steady state to find a new interesting relation [13, 21–23]. The change in the energy of the global bipartite system due to the evolution of system Y is[4]

$$\dot{E}^Y \equiv \sum_{x,y<y'} J^x_{y\to y'} \left[E_{(x,y')} - E_{(x,y)}\right], \tag{46}$$

where $E_{(x,y)}$ is the energy of state (x, y). As in (18), this energy change can be split into heat and chemical work:

$$\dot{E}^Y = \dot{Q}^Y + \dot{W}^Y_{\rm chem}, \tag{47}$$

[4] In Refs. [13, 21, 23], this quantity is denoted by $\dot{W}^{Y\to X}$ and interpreted as *transduced work* from system Y to X.

where the heat is related to the entropy change in the environment as $T\dot{S}^Y_{\text{env}} = -\dot{Q}^Y$. The same argument holds for the energy change due to the evolution of system X: $\dot{E}^X = \dot{W}^X_{\text{chem}} - T\dot{S}^X_{\text{env}}$. In the steady state, the average energy is constant, hence, $\dot{E}^X + \dot{E}^Y = 0$. Then, one can express the entropy changes in terms of the chemical works and $\dot{E}^Y$: $T\dot{S}^X_{\text{env}} = \dot{W}^X_{\text{chem}} + \dot{E}^Y$ and $T\dot{S}^Y_{\text{env}} = \dot{W}^Y_{\text{chem}} - \dot{E}^Y$. Introducing these expressions into the two inequalities (45), one obtains [13, 21]

$$\dot{W}^Y_{\text{chem}} \geq \dot{E}^Y + T\dot{I}^Y \geq -\dot{W}^X_{\text{chem}}. \tag{48}$$

The local second laws (45) and its particularization to isothermal machines (48) allow one to analyze separately the performance of each subsystem. They have been applied to coupled quantum dots [16], Brownian particles [19], sensors [24], chemical networks [25, 26], biological motors [27, 28], and other stochastic systems [13, 21]. Here we illustrate the theory with a novel application to nanomachines that couple electron transport and the oscillations of carbon nanotubes.

5 A Case Study: Nanotube Oscillations Induced by Electron Transport

In recent decades, several experimental realizations of nanoresonators activated by photons or electric transport have been implemented. These experimental setups are relevant in various applications including information processing [29], sensors [30, 31], and entanglement manipulation [32–34]. In the context of stochastic and quantum thermodynamics, nanoresonators have been used to devise nanomotors, where electron transport activates coherent oscillations in a carbon nanotube [35], as well as refrigerator engines [36], where the electron transport damps the motion of the oscillator, consequently cooling the carbon nanotube. Both regimes are autonomous and have been analyzed using stochastic thermodynamics without involving the concept of information [37–39].

In most implementations, the nanoresonator's electrical conductivity depends on the position of the oscillator. Hence, the oscillation modulates the electron current, which at the same time drives the oscillator, creating a non-linear back action between electron transport and mechanics.

Here, we study an experimental implementation based on fully suspended carbon nanotubes [41]. The left panel of Fig. 4 shows a micrograph of a typical device where a single carbon nanotube is suspended between two source-drain electrodes. Underneath the carbon nanotube, an array of five gate electrodes is used to control the transport properties of this device. With an appropriate choice of gate voltages, the five gates electrodes create an electrostatic potential along the nanotube defining a quantum dot [37, 42]. At sufficiently low temperatures, quantum dots display discrete energy transitions, and individual electron transport can be observed. The quantum dot population is thus a random variable that takes only two values, $p = 0, 1$. Apply-

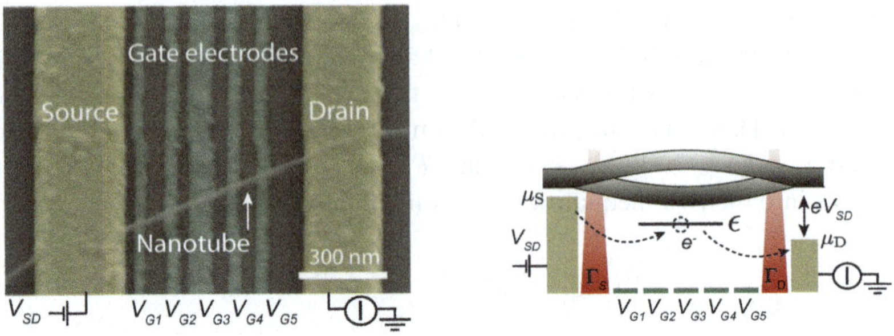

Fig. 4 The carbon nanotube resonator. *Left:* Micrograph of the device used in Ref. [40]. The nanotube (thin straight line) lies over two gold electrodes, source and drain. Below the nanotube, five gate electrodes ($V_{G1} - V_{G5}$) create a confining electrostatic potential inducing a quantum dot within the nanotube. *Right:* Scheme of the device. The electrochemical potential of the quantum dot ϵ changes with the amplitude of the oscillation. The electrons in the electrodes are distributed according to Fermi distributions with electrochemical potentials μ_{S} (source) and μ_{D} (drain), represented as gold rectangles. The voltage between the source-drain electrodes is $V_{\mathrm{SD}} = (\mu_{\mathrm{S}} - \mu_{\mathrm{D}})/e$, where e is the elementary charge. Electrons tunnel between the dot and the source-drain electrodes through potential barriers. They are represented by the two red trapezoids, indicating that the corresponding tunneling rates, $\Gamma_{\mathrm{S,D}}(\epsilon)$, depend on ϵ and, in turn, on the position of the nanotube

ing a bias voltage between the source and drain electrodes, $eV_{\mathrm{SD}} = \mu_{\mathrm{S}} - \mu_{\mathrm{D}}$, where e is the elementary charge, creates a measurable current that is proportional to the electron tunneling rates to the source and drain electrodes, Γ_S and Γ_D, respectively.

5.1 Dynamics

In this setup, mechanical oscillations are described by the amplitude x of the fundamental vibration mode of the nanotube and its velocity v. The evolution of the pair $(x(t), v(t))$ obeys the underdamped Langevin equation [43]

$$\begin{aligned} m\dot{v}(t) &= -m\omega^2 x(t) - m\frac{\omega}{\mathsf{Q}}v(t) - gp(t) + \xi(t), \\ \dot{x}(t) &= v(t). \end{aligned} \tag{49}$$

Here ω and m are the frequency and mass of the vibration mode, respectively. The friction coefficient $\gamma \equiv m\omega/\mathsf{Q}$ is given in terms of the quality factor Q, which approximately equals the number of periods for an oscillation to decay due to friction. The term $-gp(t)$ represents the force experienced by the electric charge confined within the nanotube due to the gate electrostatic field, g being a coupling constant. The Gaussian white noise $\xi(t)$ accounts for thermal fluctuations. Its average vanishes and the correlation reads $\langle \xi(t)\xi(t')\rangle = 2\gamma k T_{\mathrm{osc}}\delta(t - t')$, where T_{osc} is the temperature of the oscillator.

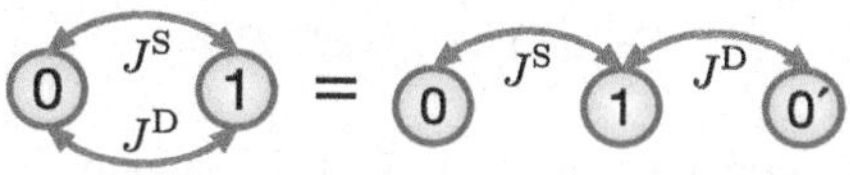

Fig. 5 Markov chain describing the occupation of the quantum dot $p(t)$. One must distinguish between the transitions due to electron exchanges with the source and drain electrodes, respectively. Each transition obeys local detailed balance and induces the corresponding current $J^{S,D}_{0\to 1}$. The resulting Markov chain is equivalent to a system with three states 0, 1, and 0′, with periodic boundary conditions that identify the two states 0 and 0′

In turn, the electrochemical potential energy of the quantum dot ϵ is capacitively coupled to the nanotube position x, due to the electrostatic field created by the gates. To be consistent with the force $-gp(t)$ in (49), this energy must read

$$\epsilon(x) = \epsilon_0 + gx. \tag{50}$$

The exchange of electrons between the dot and the electrodes is due to tunneling through potential barriers, as sketched in the right panel of Fig. 4. The barriers are characterized by tunneling rates $\Gamma_{S,D}(\epsilon)$, which may depend on the dot energy ϵ [44]. The subindex denotes respectively the corresponding electrode, source (S) or drain (D). The occupation number $p(t)$ is a Markov chain with two states, 0 or 1, and the transitions between these two states are due to the exchange of electrons between the dot and the electrodes [12]. To preserve local detailed balance, it is necessary to distinguish between transitions due to the exchange of electrons with each electrode, as shown in Fig. 5. The transition rates from $p = 0$ to $p = 1$ are $\Gamma^{S,D}_{0\to 1} = \Gamma_{S,D}(\epsilon) f_{S,D}(\epsilon)$, where $f_{S,D}(\epsilon) = 1/\left[1 + e^{\beta_{dot}(\epsilon-\mu_{S,D})}\right]$ is the Fermi function of each electrode at inverse temperature $\beta_{dot} = 1/(kT_{dot})$ and chemical potential $\mu_{S,D}$. Notice that the temperature T_{dot} of the electrodes[5] can be different from the oscillator temperature, T_{osc}. On the other hand, if the dot is occupied, the electron leaves it at a rate $\Gamma^{S,D}_{1\to 0} = \Gamma_{S,D}(\epsilon)\left[1 - f_{S,D}(\epsilon)\right]$. These transition rates obey detailed balance:

$$\frac{\Gamma^{S,D}_{0\to 1}}{\Gamma^{S,D}_{1\to 0}} = \frac{f_{S,D}(\epsilon)}{1 - f_{S,D}(\epsilon)} = e^{-\beta_{dot}(\epsilon-\mu_{S,D})}. \tag{51}$$

The system exhibits a rich phenomenology, including self-oscillations sustained by electron currents [43], if the barriers and the tunneling rates $\Gamma_{S,D}(\epsilon)$ depend on the energy. In fact, in a realistic setup, the resulting barriers are always inhomogeneous because they are very sensitive to the geometry and electrostatic properties of the device, such as local potential disorder. In a typical dot configuration, they can be parametrized as [44]

---

[5] We denote by T_{dot} the temperature of the electrodes to be consistent with our notation for entropy production in the environment of each subsystem: T_{dot} is the temperature of the environment of the dot, which consists of the two electrodes. The dot itself is in most applications out of equilibrium and its temperature (which could be defined through the ratio between the probability of being empty and occupied) has no clear physical meaning.

$$\Gamma_{S,D}(\epsilon) = \gamma_{S,D} e^{\alpha_{S,D}\epsilon/g} \tag{52}$$

where $\gamma_{S,D}$ and $\alpha_{S,D}$ are constant parameters of the device.

One can also define two probability currents, depending on x and v:

$$J^{S,D}_{0\to 1}(x, v; t) = -J^{S,D}_{1\to 0}(x, v; t) = \rho(x, v, 0; t)\Gamma^{S,D}_{0\to 1}(x) - \rho(x, v, 1; t)\Gamma^{S,D}_{1\to 0}(x) \tag{53}$$

where we have written explicitly the dependence of the transition rates on x through the energy of the dot: $\epsilon(x) = \epsilon_0 + gx$.

The dynamics of the whole system is governed by the Langevin equation (49) together with the stochastic jumps of the dot occupation $p(t)$. They induce an evolution equation for the probability distribution $\rho(x, v, p; t)$, which is a combination of a Fokker-Planck and a master equation:

$$\frac{\partial \rho(x, v, p; t)}{\partial t} = -\frac{\partial v\rho}{\partial x} - \frac{\partial (F/m)\rho}{\partial v} + \frac{\omega k T_{\rm osc}}{m\mathsf{Q}}\frac{\partial^2 \rho}{\partial v^2} + (2p-1)\left[J^{S}_{0\to 1} + J^{D}_{0\to 1}\right], \tag{54}$$

where $F(x, v, p; t) = -m\omega^2 x - m\omega v/\mathsf{Q} - gp$ is the deterministic force acting on the nanotube.

In terms of probability currents, the equation can be written as

$$\frac{\partial \rho(x, v, p; t)}{\partial t} = -\nabla\cdot\left[\mathbf{J}^{\rm rev}_p(x, v; t) + \mathbf{J}^{\rm irrev}_p(x, v; t)\right] + (2p-1)\left[J^{S}_{0\to 1} + J^{D}_{0\to 1}\right], \tag{55}$$

where the gradient is defined on the plane (x, v), i.e., $\nabla \equiv (\partial_x, \partial_v)$. In this expression, we have split the probability current associated with the nanotube motion into a conservative or reversible term and an irreversible contribution [45]:

$$\begin{aligned} \mathbf{J}^{\rm rev}_p(x, v; t) &\equiv \frac{1}{m}\begin{pmatrix} mv \\ -m\omega^2 x - gp \end{pmatrix}\rho(x, v, p; t) \\ \mathbf{J}^{\rm irrev}_p(x, v; t) &\equiv -\frac{\omega}{m\mathsf{Q}}\begin{pmatrix} 0 \\ \partial_v\left[mv^2/2 + kT_{\rm osc}\ln\rho\right]\end{pmatrix}\rho(x, v, p; t). \end{aligned} \tag{56}$$

5.2 *Thermodynamics and Information Flows*

Here we proceed as in Sect. 3.1: we calculate the entropy production rate, $\dot{S}_{\rm prod}(t)$, as the sum of the change in the Shannon entropy of the system $\dot{S}(t)$ plus the change in the thermodynamic entropy of the environment $\dot{S}_{\rm env}(t)$. The contribution of the dot occupation to $\dot{S}_{\rm env}(t)$, i.e., the rate of change in the entropy of the electrodes, can be expressed as in (43):

$$\dot{S}^{\text{dot}}_{\text{env}}(t) = k \int dxdv \left[J^{\text{S}}_{0\to1} \ln \frac{\Gamma^{\text{S}}_{0\to1}}{\Gamma^{\text{S}}_{1\to0}} + J^{\text{D}}_{0\to1} \ln \frac{\Gamma^{\text{D}}_{0\to1}}{\Gamma^{\text{D}}_{1\to0}} \right]$$
$$= \frac{1}{T_{\text{dot}}} \int dxdv \left[J^{\text{S}}_{0\to1}(x, v; t)(\mu_S - \epsilon(x)) + J^{\text{D}}_{0\to1}(x, v; t)(\mu_D - \epsilon(x)) \right]. \tag{57}$$

To proceed, we have to quantify the change in the entropy of the environment due to mechanical oscillations. A way to do this is to calculate the variation of energy in the nanotube and identify the heat and work terms. Using the Fokker-Planck equation (54) and integrating by parts, one can write the change of energy as

$$\dot{E}(t) \equiv \frac{d}{dt}\left\langle \frac{mv^2}{2} + \frac{m\omega^2 x^2}{2} + p\epsilon(x) \right\rangle$$
$$= \sum_{p=0,1} \int dxdv \left[\frac{mv^2}{2} + \frac{m\omega^2 x^2}{2} + p\epsilon(x) \right] \frac{\partial \rho(x, v, p; t)}{\partial t}$$
$$= \frac{\omega}{\mathsf{Q}} \left[kT_{\text{osc}} - m\langle v(t)^2 \rangle \right] + \int dxdv\, \epsilon(x) \left[J^{\text{S}}_{0\to1}(x, v; t) + J^{\text{D}}_{0\to1}(x.v; t) \right]$$
$$= \dot{Q}_{\text{osc}}(t) + \dot{Q}_{\text{dot}}(t) + \dot{W}_{\text{chem}}(t) \tag{58}$$

where the heat from the thermal bath of the oscillator is [45, 46]

$$\dot{Q}_{\text{osc}}(t) = -T_{\text{osc}} \dot{S}^{\text{osc}}_{\text{env}}(t) = \frac{\omega}{\mathsf{Q}} \left[kT_{\text{osc}} - m\langle v(t)^2 \rangle \right]. \tag{59}$$

As in (18), the contribution of the electron current to the total energy change in (58) is the sum of the heat transferred from the electrodes $\dot{Q}_{\text{dot}}(t) = -T_{\text{dot}} \dot{S}^{\text{dot}}_{\text{env}}(t)$, with $\dot{S}^{\text{dot}}_{\text{env}}(t)$ given by (57), and the chemical work

$$\dot{W}_{\text{chem}}(t) = \int dxdv \left[J^{\text{S}}_{0\to1}(x, v; t)\mu_{\text{S}} + J^{\text{D}}_{0\to1}(x, v; t)\mu_{\text{D}} \right]. \tag{60}$$

Now, we have all the elements to find the analogous to the local second laws (45). We first consider the total Shannon entropy of the system

$$S(x, v, p; t) = -k \sum_{p=0,1} \int dxdv\, \rho(x, v, p; t) \ln \rho(x, v, p; t)$$
$$= S_{\text{osc}}(t) + S_{\text{dot}}(t) - I(x, v; p) \tag{61}$$

where $S_{\text{osc}}(t)$ and $S_{\text{dot}}(t)$ are the Shannon entropies of each subsystem and the mutual information reads

$$I(x, v; p) = k \sum_{p=0,1} \int dxdv\, \rho(x, v, p; t) \ln \frac{\rho(x, v, p; t)}{\rho(x, v; t)\rho(p; t)}. \tag{62}$$

The time derivative of the Shannon entropy can be split into a contribution from the dot and another from the oscillator

$$\dot{S}(x, v, p; t) = \dot{S}_{\mathrm{osc}}(t) + \dot{S}_{\mathrm{dot}}(t) - \dot{I}^{\mathrm{osc}}(t) - \dot{I}^{\mathrm{dot}}(t) \tag{63}$$

with

$$\dot{I}^{\mathrm{osc}}(t) \equiv -k \sum_{p=0,1} \int dx dv\, \nabla \cdot \left[\mathbf{J}_p^{\mathrm{rev}}(x, v; t) + \mathbf{J}_p^{\mathrm{irrev}}(x, v; t)\right] \ln \frac{\rho(x, v, p; t)}{\rho(x, v; t)} \tag{64}$$

and

$$\dot{I}^{\mathrm{dot}}(t) \equiv k \sum_{p=0,1} \int dx dv\, (2p-1) \left[J^{\mathrm{S}}_{0\to 1}(x, v; t) + J^{\mathrm{D}}_{0\to 1}(x, v; t)\right] \ln \frac{\rho(x, v, p; t)}{\rho(p; t)}. \tag{65}$$

Finally, we can sum up the Shannon entropy and the change in the entropy of the respective environments as in Sect. 4.3. After some cumbersome but straightforward algebra, we obtain the two local second laws:

$$\dot{S}^{\mathrm{osc}}_{\mathrm{prod}}(t) \equiv \dot{S}_{\mathrm{osc}}(t) - \dot{I}^{\mathrm{osc}}(t) + \dot{S}^{\mathrm{osc}}_{\mathrm{env}}(t) = \frac{Qm}{\omega T_{\mathrm{osc}}} \left\langle \left[\frac{\mathbf{J}_p^{\mathrm{irrev}}(x, v; t)}{\rho(x, v, p; t)}\right]^2 \right\rangle \geq 0 \tag{66}$$

and

$$\begin{aligned} \dot{S}^{\mathrm{dot}}_{\mathrm{prod}}(t) &\equiv \dot{S}_{\mathrm{dot}}(t) - \dot{I}^{\mathrm{dot}}(t) + \dot{S}^{\mathrm{dot}}_{\mathrm{env}}(t) \\ &= k \int dx dv \sum_{\alpha=\mathrm{S,D}} J^{\alpha}_{0\to 1}(x, v; t) \ln \frac{\rho(x, v, 0; t)\Gamma^{\alpha}_{0\to 1}}{\rho(x, v, 1; t)\Gamma^{\alpha}_{1\to 0}} \geq 0, \end{aligned} \tag{67}$$

which is a special case of (44). The sum of the two yields the second law for the global system:

$$\dot{S}_{\mathrm{prod}}(t) = \dot{S}_{\mathrm{osc}}(t) + \dot{S}_{\mathrm{dot}}(t) - \dot{I}(t) - \frac{\dot{Q}_{\mathrm{osc}}(t)}{T_{\mathrm{osc}}} - \frac{\dot{Q}_{\mathrm{dot}}(t)}{T_{\mathrm{dot}}} \geq 0. \tag{68}$$

Finally, we particularize these results when the device works in the stationary regime. The marginal entropies $S_{\mathrm{osc}}(t)$ and $S_{\mathrm{dot}}(t)$ are constant and the information flows obey $\dot{I}^{\mathrm{dot}} + \dot{I}^{\mathrm{osc}} = 0$. Hence, the local second laws (66) and (67) read

$$\begin{aligned} \dot{S}^{\mathrm{osc}}_{\mathrm{prod}} &= \dot{S}^{\mathrm{osc}}_{\mathrm{env}} - \dot{I}^{\mathrm{osc}} = -\frac{\dot{Q}_{\mathrm{osc}}}{T_{\mathrm{osc}}} - \dot{I}^{\mathrm{osc}} \geq 0 \\ \dot{S}^{\mathrm{dot}}_{\mathrm{prod}} &= \dot{S}^{\mathrm{dot}}_{\mathrm{env}} - \dot{I}^{\mathrm{dot}} = -\frac{\dot{Q}_{\mathrm{dot}}}{T_{\mathrm{dot}}} + \dot{I}^{\mathrm{osc}} \geq 0. \end{aligned} \tag{69}$$

If we use the conservation of energy, $\dot{E}(t) = 0$ in (58), $\dot{Q}_{\rm osc} + \dot{Q}_{\rm dot} + \dot{W}_{\rm chem} = 0$, we obtain the following constraints for the performance of the device:

$$\begin{aligned}\dot{S}^{\rm osc}_{\rm prod} &= -\frac{\dot{Q}_{\rm osc}}{T_{\rm osc}} - \dot{I}^{\rm osc} \geq 0 \\ \dot{S}^{\rm dot}_{\rm prod} &= \frac{\dot{Q}_{\rm osc} + \dot{W}_{\rm chem}}{T_{\rm dot}} + \dot{I}^{\rm osc} \geq 0.\end{aligned} \tag{70}$$

In the stationary regime, the chemical work $\dot{W}_{\rm chem}$, defined in Eq.(60), adopts the familiar form:

$$\dot{W}_{\rm chem} = eV_{\rm SD}J, \tag{71}$$

where e is the elementary charge and

$$J \equiv \int dxdv J^{\rm S}_{0\to 1}(x, v; t) = -\int dxdv J^{\rm D}_{0\to 1}(x, v; t) \tag{72}$$

is the net electron current from the source to the drain electrode. Finally, the heat from the electrodes also simplifies in the stationary regime:

$$\dot{Q}_{\rm dot} = -T_{\rm dot}\dot{S}^{\rm dot}_{\rm env} = -\dot{W}_{\rm chem} + \int dxdv \left[J^{\rm S}_{0\to 1}(x, v; t) + J^{\rm D}_{0\to 1}(x, v; t)\right]\epsilon(x). \tag{73}$$

The global second law is the sum of the inequalities (70), which can be written as

$$\dot{W}_{\rm chem} + \dot{Q}_{\rm osc}\left[1 - \frac{T_{\rm dot}}{T_{\rm osc}}\right] \geq 0 \tag{74}$$

and establishes the main limitations of the device and its possible operating modes.

In the next subsections, we study two modes of operation in the stationary regime, whose energetics are depicted in Fig. 6. In this figure, the arrows indicate the direction of the energy transfer. First, we analyze the device acting as an electron pump that moves electrons against the voltage: $\dot{W}_{\rm chem} < 0$. The energy to perform this work is extracted from a hot bath and part of it is dissipated as heat in a cold bath, to comply with the second law of thermodynamics (74). More precisely, if $\dot{W}_{\rm chem} < 0$, then (74) is satisfied only if $T_{\rm osc} < T_{\rm dot}$ and $\dot{Q}_{\rm osc} < 0$ or if $T_{\rm osc} > T_{\rm dot}$ and $\dot{Q}_{\rm osc} > 0$. Here we will only discuss the first case, where the entropy increases in the oscillator bath and decreases in the electrodes. The second mode of operation that we will consider below is the case where the electron current performs work, $\dot{W}_{\rm chem} > 0$, and this work is used to pump heat from the thermal bath of the nanotube to the electrodes against the temperature gradient, i.e., $\dot{Q}_{\rm osc} > 0$ and $\dot{Q}_{\rm dot} < 0$ when $T_{\rm osc} \leq T_{\rm dot}$. Notice that the information flow is a necessary ingredient in these two regimes of operation.

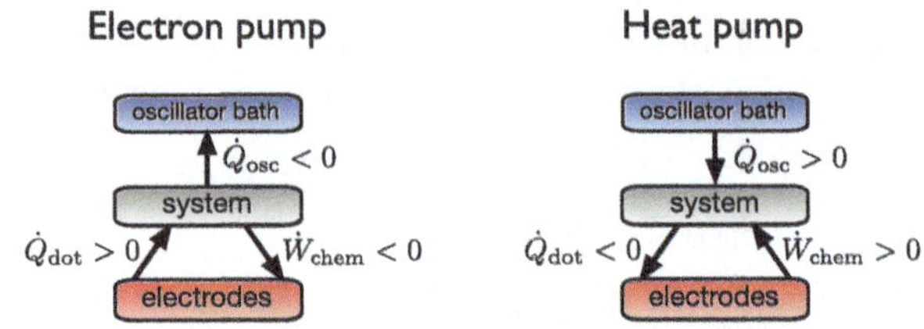

Fig. 6 Energetics of the devices in the two regimes discussed in the text. The arrows indicate the direction of the energy transfer. *Left:* the electron pump transfers electrons against the voltage performing work, $\dot{W}_{chem} < 0$. The energy is extracted from the electrodes at temperature T_{dot} ($\dot{Q}_{dot} > 0$) and part of this energy is dissipated to the bath of the oscillator ($\dot{Q}_{osc} < 0$) at lower temperature, $T_{osc} < T_{dot}$, to compensate the decrease of entropy in the electrodes. *Right:* a heat pump that transfers energy from the thermal bath of the oscillator ($\dot{Q}_{osc} > 0$) to the electrodes ($\dot{Q}_{dot} < 0$) against the temperature difference, i.e., when the oscillator bath is colder than the electrodes, $T_{osc} < T_{dot}$. To achieve this regime, the electron current must provide energy performing work, i.e. $\dot{W}_{chem} > 0$

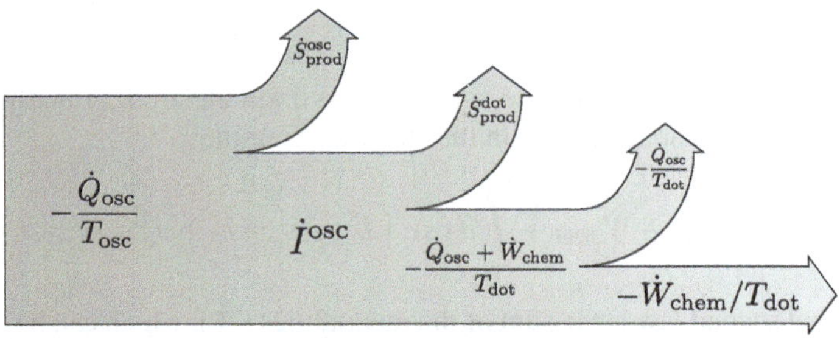

Fig. 7 Sankey diagram of the electron pump. In this operation regime, heat is dissipated by the oscillator, $\dot{Q}_{osc} < 0$, and the dot pumps electrons against the voltage, $\dot{W}_{chem} = eV_{SD}J < 0$. The energy comes from the heat transferred from the electrodes to the dot, $\dot{Q}_{dot} = -\dot{Q}_{osc} - \dot{W}_{chem} > 0$. To achieve this regime it is necessary that the environment of the oscillator is colder than the electrodes, $T_{dot} > T_{osc}$

5.3 *Electron Pump*

In the electron pump regime, the device converts energy from the oscillations into a net current of electrons between the source and drain electrodes, even against an applied bias voltage. We further assume that $T_{osc} < T_{dot}$. The energetics is depicted in the left panel of Fig. 6 and the corresponding Sankey diagram is shown in Fig. 7, which is a special case of the diagram in Fig. 3. Here, the nanotube plays the role of system Y and the dot is system X. Hence, the information flow $\dot{I}^{osc} = -\dot{I}^{dot}$ is positive and the nanotube can be seen as a Maxwell demon operating on the quantum dot. In Fig. 7, we have added a last stage to the diagram reflecting that the useful output is the chemical work.

From the global second law (74), it is straightforward to bound the efficiency of the pump:

$$\eta \equiv \frac{-\dot{W}_{\text{chem}}}{\dot{Q}_{\text{dot}}} = \frac{\dot{Q}_{\text{dot}} + \dot{Q}_{\text{osc}}}{\dot{Q}_{\text{dot}}} \leq \eta_{\text{Carnot}} \tag{75}$$

where $\eta_{\text{Carnot}} \equiv 1 - T_{\text{osc}}/T_{\text{dot}}$ is the Carnot efficiency. The equality is reached, as expected, when the total entropy production vanishes. Notice that the last "leak" in the Sankey diagram is not an entropy production but a loss due to the fact that not all the decrease in the entropy of the environment of the dot, $\dot{Q}_{\text{dot}}/T_{\text{dot}} = -(\dot{Q}_{\text{osc}} + \dot{W}_{\text{chem}})/T_{\text{dot}}$, is used to perform chemical work.

As already mentioned, the information flow suggests that the ocsillator acts as a demon to the dot. This interpretation can be made more precise by analyzing the dynamics of the device. Figure 8 sketches the operation of the electron pump for the case of zero bias $V_{\text{SD}} = 0$, i.e. when the electrochemical potentials of the source and drain electrodes are the same, $\mu_{\text{S}} = \mu_{\text{D}}$. Recall that the amplitude of the nanotube behaves as a particle in a harmonic potential $m\omega^2 x^2/2 + gpx$, which depends on whether the dot is filled ($p = 1$) or empty ($p = 0$). The equilibrium position of these two potentials is, respectively, $x_0 = 0$ and $x_1 = -g/(m\omega^2)$, for which the electrochemical potential of the dot $\epsilon(x) = \epsilon_0 + gx$ is, respectively, $\epsilon(x_0) = \epsilon_0$ and $\epsilon(x_1) = \epsilon_0 - g^2/(m\omega^2) \equiv \epsilon_1$. The pump regime is achieved when these values are close to the electrochemical potential of the source and drain electrodes. The figure shows the case $\epsilon_0 = \mu_{\text{S}} = \mu_{\text{D}}$ and $\epsilon_0 - \epsilon_1 = g^2/(m\omega^2) \lesssim kT_{\text{dot}}$. Another necessary condition for the pump to work is that at least one of the tunnel barriers depends on the energy, $\Gamma_{\text{S,D}}(\epsilon)$. In the figure, we have assumed that the left barrier (source) is homogeneous, $\alpha_{\text{S}} = 0$, and that the right barrier (drain) narrows when the energy of the dot increases, i.e. $\alpha_{\text{D}} > 0$ and $\Gamma_{\text{D}}(\epsilon)$ is an increasing function of ϵ, according to equation (52).

When the dot is empty, the oscillator's equilibrium position is x_0 and the electrochemical potential of the dot is ϵ_0. Since $\Gamma_{\text{S}}(\epsilon_0) < \Gamma_{\text{D}}(\epsilon_0)$, the dot is filled by electrons coming, preferentially, from the drain electrode. Once the dot is filled, the oscillator moves to x_1 and shifts the electrochemical potential of the dot to ϵ_1. In this case, $\Gamma_{\text{S}}(\epsilon_0) > \Gamma_{\text{D}}(\epsilon_0)$ and the electron leaves the dot preferentially from the source electrode. Hence, the device exhibits a net current of electrons from the drain to the source, and this current persists against moderate values of the bias voltage $V_{\text{SD}} = \mu_{\text{S}} - \mu_{\text{D}} \geq 0$. Notice that the oscillator acts as a Maxwell demon: it "measures" $p = 0, 1$, i.e. detects whether the dot is filled or not, and changes the electrochemical potential of the dot in a way that facilitates the transfer of electrons from the right electrode when the dot is empty and to the left electrode when the dot is filled. We illustrate this operation in the right panel of Fig. 8, depicting the demon opening and closing trapdoors that connect the dot with the source-drain electrodes in a manner that induces a current from right to left.

We have conducted numerical simulations of the device in the electron pump regime. In Fig. 9, we show the four quantities appearing in the Sankey diagram of Fig. 7: the heat flows $\dot{Q}_{\text{dot}}$ and $\dot{Q}_{\text{osc}}$, the chemical work $\dot{W}_{\text{chem}}$, and the information flow, $\dot{I}^{\text{dot}} = -\dot{I}^{\text{osc}}$ for several values of $\Gamma \equiv \gamma_{\text{S,D}}$ and bias voltage V_{SD}.

The maximum power is reached when $\Gamma \simeq \omega$, since in this case the demon operation, namely, the oscillations, and the electron transfer occur on similar time scales.

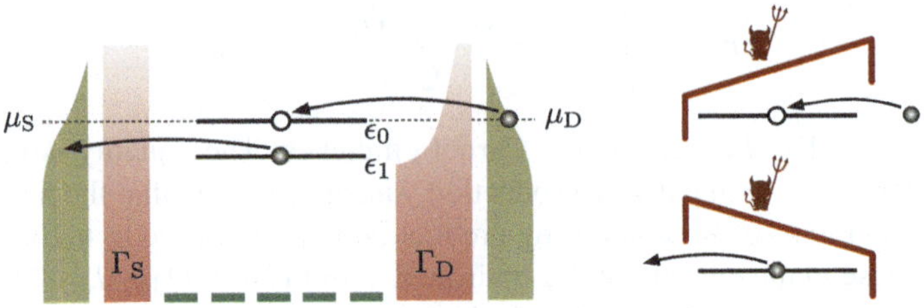

Fig. 8 *Left*: Sketch of the electron pump mechanism at zero bias voltage, $V_{SD} = 0$ (i.e., for $\mu_S = \mu_D$). This mechanism is based on the inhomogeneity of the tunnel barriers $\Gamma_{S,D}(\epsilon)$ and the shift of the equilibrium position of the oscillator when the dot is either filled ($p = 1$) or empty ($p = 0$). The electrochemical potential of the dot $\epsilon(x) = \epsilon_0 + gx$ is tuned to match the Fermi levels of the source and drain electrodes when the oscillator is empty and in its equilibrium position $x_0 = 0$, i.e., $\epsilon(x_0) = \epsilon_0 = \mu_S = \mu_D$. When the dot is filled with an electron ($p = 1$), the equilibrium position of the oscillator is $x_1 = -g/(m\omega^2)$. At this position, the electrochemical potential of the dot is $\epsilon_1 = \epsilon_0 - g^2/(m\omega^2)$. If T_{osc} is low enough and the tunnel barriers satisfy $\Gamma_S(\epsilon_0) < \Gamma_D(\epsilon_0)$ and $\Gamma_S(\epsilon_1) > \Gamma_D(\epsilon_1)$, then the device exhibits a net current from the drain to the source. *Right*: The operation of the oscillator is sketched as a Maxwell demon that operates two trapdoors connecting the dot with the source and drain electrodes. The demon opens the right trapdoor and closes the left trapdoor when the dot is empty, whereas it opens the left trapdoor and closes the right one when the dot is filled. The operation generates a current from right to left

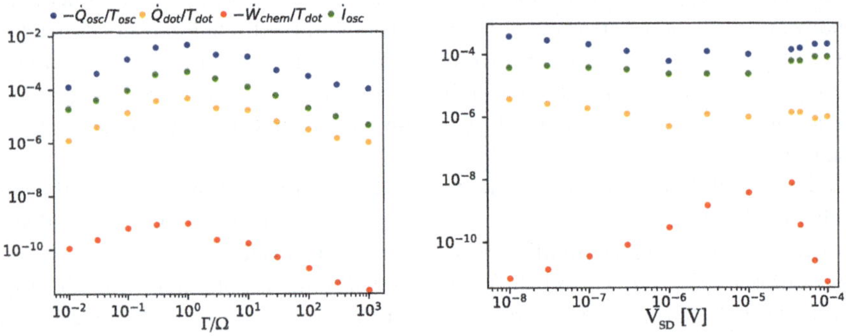

Fig. 9 The four entropy flows in the Sankey diagram of Fig. 7: $-\dot{Q}_{osc}/T_{osc}$ (blue points), $\dot{I}^{dot}$ (green), $\dot{Q}_{dot}/T_{dot}$ (orange), and $-\dot{W}_{chem}/T_{dot}$ (red), in units of Boltzmann constant per oscillation ($2\pi k/\omega$). *Left*: Flows as a function of the tunnel rates $\gamma_S = \gamma_D = \Gamma$, for $V_{SD} = 0.1\ \mu$V. *Right*: Flows as a function of V_{SD} with $\Gamma/\omega = 100$. In both panels $T_{dot}/T_{osc} = 10$, $\mu_S = \epsilon_0 = 0$, $kT_{osc} = 0.01$ eV, $\omega/2\pi = 270$ MHz, $g = 0.01$ eV/nm, $m = 2 \times 10^{-22}$ kg, Q $= 1000$, $\alpha_S = 0$, and $\alpha_D = 4$ nm^{-1} (Color online)

We observe that the chemical work is small compared with the heat dissipated, $\dot{Q}_{dot}, \dot{Q}_{osc} \gg \dot{W}_{chem}$, that is, the electron pump mechanism is very inefficient from the point of view of thermodynamics, considering realistic device parameters. Notice however that the entropy production is relatively small compared to the last step in the Sankey diagram which is due to the different temperatures in the environments and limits the performance of the pump below the Carnot efficiency.

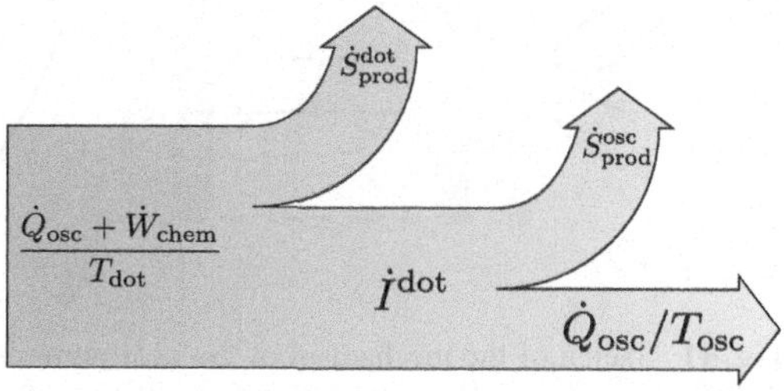

Fig. 10 Sankey diagram of a heat pump. In this case, $\dot{Q}_{osc} > 0$ is the heat extracted from the cold bath, $T_{osc} < T_{dot}$, and the electron current performs a work $\dot{W}_{chem} = eV_{SD}J > 0$

In the right panel of Fig. 9, we show the entropic flows as a function of the bias voltage V_{SD}. When the bias voltage is low, the electrons are pumped without resistance, performing no chemical work. If we increase the bias, the system starts performing work until $V_{SD} \sim 10^{-4}$ V, where the bias voltage completely counteracts the electron pump, reversing the direction of the current. By converting V_{SD} to a temperature, we find that thermal and other sources of fluctuations must be reduced below approximately 110 mK in order for the pump current to be measurable in an experiment, a condition that is achieved in standard setups.

5.4 *Heat Pump*

Our second example is a heat pump, whose energetics is represented in the right panel of Fig. 6. In this regime, the direction of the electron current is the one given by V_{SD} and $T_{osc} < T_{dot}$. Hence, the work is positive, $\dot{W}_{chem} > 0$, and this energy is used to pump heat from the cold thermal bath affecting the oscillator ($\dot{Q}_{osc} > 0$) to the hot thermal bath, namely, the electrodes ($\dot{Q}_{dot} < 0$). This behavior can be seen as an *anti-Joule* effect since the current cools down the nanotube using its motion.

The corresponding entropic Sankey diagram is depicted in Fig. 10. As in the case of the electron pump, the global second law imposes an upper bound to the efficiency of the machine [47]:

$$\eta \equiv \frac{\dot{Q}_{osc}}{\dot{W}_{chem}} \leq \frac{T_{osc}}{T_{dot} - T_{osc}} \tag{76}$$

which diverges when the two temperatures are equal. The refrigeration power is also bound by

$$\dot{Q}_{osc} < \frac{kT_{osc}}{\mathsf{Q}}\omega, \tag{77}$$

as can be deduced from (59). This inequality is more general and is verified by systems in which energy is extracted from a bath by means of a mechanical degree of freedom. The inequality is due to the fact that energy is extracted when the velocity fluctuations $\langle v^2 \rangle$ are smaller than the fluctuations at equilibrium kT/m [46].

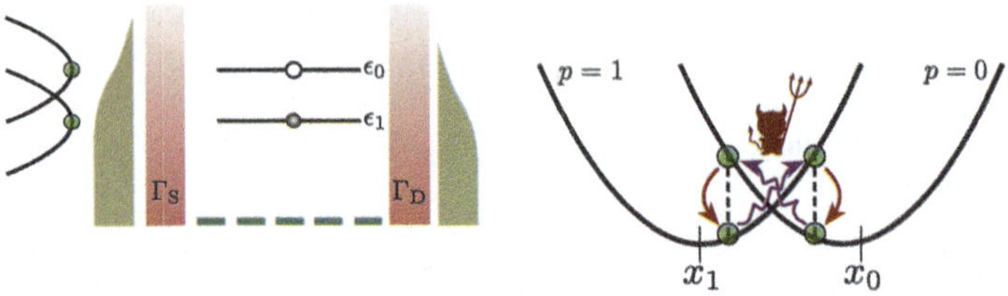

Fig. 11 Sketch of the mechanism of the heat pump. *Left*: When the dot is empty, the oscillator fluctuates around x_0. However, a significant thermal fluctuation can push the oscillator downward, thereby increasing the likelihood of the dot being filled. When this happens, the potential that the oscillator feels is shifted downward, and the oscillator transfers energy to the dot. A similar effect happens when a fluctuation drives the oscillator upwards and the dot loses its electron. *Right*: A Maxwell demon extracting energy from the environment of an oscillator acts in a similar way. When a thermal fluctuation drives the oscillator (green circle) far from the corresponding equilibrium position, i.e., to a point with high potential energy, the demon shifts the potential reducing the energy of the oscillator. The demon is extracting the energy that the environment transfers to the oscillator through the fluctuations. The purple zigzag arrows indicate the motion of the oscillator due to fluctuations, and the red arrows indicate the change in energy due to the switch in the potential carried out by the demon

The positive information flow now corresponds to the quantum dot, $\dot{I}^{\text{dot}} = -\dot{I}^{\text{osc}} > 0$, indicating that, in this regime, the dot functions as a Maxwell demon operating on the oscillator. In Fig. 11, we present an interpretation of the device's operation. In this case, it is more illustrative to consider how energy can be extracted from the thermal bath of an oscillator through feedback. The procedure is illustrated in the right panel of Fig. 11. In the device, when the quantum dot is empty ($p = 0$), the oscillator moves around the equilibrium position x_0 thereby setting the electrochemical potential of the dot to ϵ_0. In this configuration, if a fluctuation pushes the oscillator downward, the probability of the dot being filled increases. Consequently, for sufficiently large fluctuations, the dot is filled with an electron when the oscillator moves downward. This corresponds to the demon switching the potential to $p = 1$, which changes the equilibrium position of the oscillator to x_1 and sets the electrochemical potential of the dot to ϵ_1. In Fig. 11, fluctuations are depicted as purple zigzag arrows and the potential switch induced by the demon is represented by red arrows. The switch reduces the energy of the system, which is absorbed by the demon. Once the potential corresponds to $p = 1$, the demon repeats the process, waiting now for a sufficiently large fluctuation to push the oscillator upwards closer to x_0.

We have performed numerical simulations of the model for equal temperatures $T_{\text{osc}} = T_{\text{dot}} = T$. In this case, the second law (76) does not provide an upper finite bound on the efficiency of the device, because a heat flow can be established with arbitrarily small work. However, the refrigeration power $\dot{Q}_{\text{osc}}$ is limited by equation (77). In Fig. 12 we depict the energy transfers: $\dot{W}_{\text{chem}}$ (red), $T\dot{I}^{\text{dot}}$ (green), $\dot{Q}_{\text{osc}}$ (blue), and the bound $kT\omega/\text{Q}$ (dashed line) as a function of the tunneling rates Γ and the temperature T. We observe that the highest entropy production occurs in the electron current, except in cases of low tunneling rates Γ.

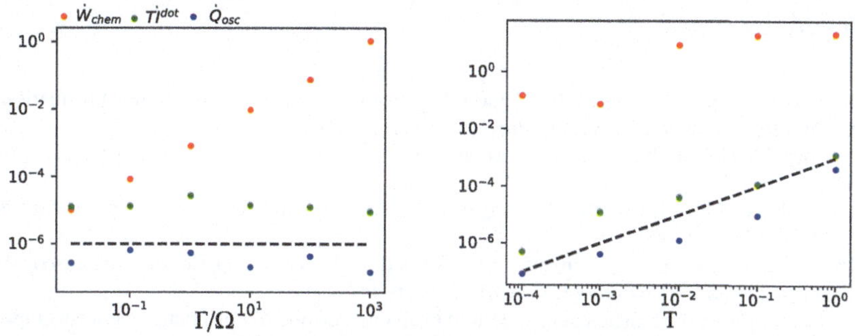

Fig. 12 Heat flow $\dot{Q}_{\mathrm{osc}}$ (blue points), chemical work $\dot{W}_{\mathrm{chem}}$ (red points), and information flow $T_{\mathrm{dot}}\dot{I}^{\mathrm{dot}}$ (green points) in eV per oscillation: *Left*: for several tunnel rates $\gamma_S = \gamma_D = \Gamma$ with $kT_{\mathrm{osc}} = kT_{\mathrm{dot}} = 0.001$ eV; *Right*: for several temperatures $T_{\mathrm{osc}} = T_{\mathrm{dot}} = T$ with $\Gamma = 100\omega$. The black dashed line is the bound (77), i.e., $kT\omega/\mathrm{Q}$. The bias voltage is $VSD = -2$ mV and the rest of the parameters are the same as in Fig. 9 (Color online)

6 Conclusions

We have provided a self-contained review of the definition and use of information flows in bipartite systems that exchange energy and particles with reservoirs. The main result is the two local second laws (45), which impose an extra constraint on the energetics of a bipartite system in a non-equilibrium steady state (NESS). As illustrated by the Sankey diagram in Fig. 3, the information flow $\dot{I}^Y = -\dot{I}^X$ quantifies the transfer of negative entropy between the two subsystems, X and Y. Moreover, it enables us to interpret the overall system as an information machine, where one of the subsystems plays the role of a Maxwell demon, measuring and operating on the other subsystem [16, 20]. This interpretation holds the potential to enhance the design of nanomachines and the understanding of biological motors.

We have illustrated these ideas by analyzing a bipartite system consisting of a quantum dot in a carbon nanotube, where the mechanical oscillations of the nanotube are coupled to electron transport. Information flows allow one to interpret the device as an information machine. This interpretation provides hints for improving the efficiency or even achieving the desired functionality. In the case of the electron pump, for instance, the inhomogeneity of the tunnel barriers is a necessary condition for pumping electrons against the bias voltage. Mutual information reveals that this inhomogeneity is necessary to create a correlation between the population of the dot and the position of the nanotube.

Acknowledgements We thank Jordan Horowitz and Jannik Ehrich for their thoughtful comments. This research was supported by grant number FQXi-IAF19-01 from the Foundational Questions Institute Fund, a donor-advised fund of Silicon Valley Community Foundation. JMRP and JT-B acknowledge financial support from Spanish Government through Grant FLUID (PID2020-113455GB-I00). NA and FF acknowledge the support from the Royal Society (URF-R1-191150), from the European Research Council (ERC) under the European Union's Horizon 2020 research and innovation programme (grant agreement number 948932), and the European Commission via the Horizon Europe project ASPECTS (Grant Agreement No. 101080167).

References

1. Leff, H.S., Rex, A.F.: Maxwell's demon 2: entropy. Classical and quantum information. In: Computing. Princeton University Press, New Jersey (2002)
2. Sagawa, T.: Thermodynamics of Information Processing in Small Systems. Springer, Springer Theses (2012)
3. Parrondo, J.M.R., Horowitz, J.M., Sagawa, T.: Thermodynamics of information. Nat. Phys. **11**(2), 131–139 (2015)
4. Parrondo, J.M.R.: Thermodynamics of information. In: Encyclopedia of Condensed Matter Physics, 2nd edn. Elsevier (2023). arXiv:2306.12447v1
5. Szilard, L.: Über die entropieverminderung in einem thermodynamischen system bei eingriffen intelligenter wesen. Zeitschrift für Physik **53**(11), 840–856 (1929). On the decrease of entropy in a thermodynamic system by the intervention of intelligent beings. Translated into English and reprinted in [1]
6. Sagawa, T., Ueda, M.: Second law of thermodynamics with discrete quantum feedback control. Phys. Rev. Lett. **100**, 080403 (2008)
7. Cover, T.M., Thomas, J.A.: Elements of Information Theory, 2nd edn. Wiley-Interscience, New York (2006)
8. Landauer, R.: Irreversibility and heat generation in the computing process. IBM J. Res. Dev. **5**(3), 183–191 (1961)
9. Bennett, C.H.: The thermodynamics of computation—a review. Int. J. Theor. Phys. **21**(12), 905–940 (1982). Reprinted in [1]
10. Esposito, M., Van den Broeck, C.: Three faces of the second law. I. Master equation formulation. Phys. Rev. E **82**, 011143 (2010)
11. Van den Broeck, C., Esposito, M.: Three faces of the second law. II. Fokker-Planck formulation. Phys. Rev. E **82**, 011144 (2010)
12. Esposito, M., Harbola, U., Mukamel, S.: Entropy fluctuation theorems in driven open systems: application to electron counting statistics. Phys. Rev. E **76**, 031132 (2007)
13. Grelier, M., Sivak, D.A., Ehrich, J.: Unlocking the potential of information flow: maximizing free-energy transduction in a model of an autonomous rotary molecular motor (2023). arXiv:2304.06690
14. Horowitz, J.M., Parrondo, J.M.R.: Designing optimal discrete-feedback thermodynamic engines. New J. Phys. **13**, 123019 (2011)
15. Horowitz, J.M., Parrondo, J.M.R.: Thermodynamic reversibility in feedback processes. Europhys. Lett. **95**, 10005 (2011)
16. Horowitz, J.M., Esposito, M.: Thermodynamics with continuous information flow. Phys. Rev. X **4**(3), 031015 (2014)
17. Barato, A.C., Seifert, U.: Stochastic thermodynamics with information reservoirs. Phys. Rev. E **90**, 042150 (2014)
18. Mandal, D., Jarzynski, C.: Work and information processing in a solvable model of Maxwell's demon. Proc. Natl. Acad. Sci. **109**(29), 11641–11645 (2012)
19. Allahverdyan, A.E., Janzing, D., Mahler, G.: Thermodynamic efficiency of information and heat flow. J. Stat. Mech.: Theory Exp. **2009**(09), P09011 (2009)
20. Hartich, D., Barato, A.C., Seifert, U.: Stochastic thermodynamics of bipartite systems: transfer entropy inequalities and a Maxwell's demon interpretation. J. Stat. Mech.: Theory Exp. **2014**(2), P02016 (2014)
21. Ehrich, J., Sivak, D.A.: Energy and information flows in autonomous systems. Front. Phys. 11 (2023)
22. Barato, A.C., Seifert, U.: Thermodynamic cost of external control. New J. Phys. **19**(7), 073021 (2017)
23. Large, S.J., Ehrich, J., Sivak, D.A.: Free-energy transduction within autonomous systems. Phys. Rev. E **103**, 022140 (2021)
24. Hartich, D., Barato, A.C., Seifert, U.: Sensory capacity: an information theoretical measure of the performance of a sensor. Phys. Rev. E **93**, 022116 (2016)

25. Penocchio, E., Avanzini, F., Esposito, M.: Information thermodynamics for deterministic chemical reaction networks. J. Chem. Phys. **157**(3), 034110 (2022)
26. Amano, S., Esposito, M., Kreidt, E., Leigh, D.A., Penocchio, E., Roberts, B.M.W.: Insights from an information thermodynamics analysis of a synthetic molecular motor. Nat. Chem. **14**(5), 530–537 (2022)
27. Lathouwers, E., Sivak, D.A.: Internal energy and information flows mediate input and output power in bipartite molecular machines. Phys. Rev. E **105**, 024136 (2022)
28. Takaki, R., Mugnai, M.L., Thirumalai, D.: Information flow, gating, and energetics in dimeric molecular motors. Proc. Natl. Acad. Sci. **119**(46), e2208083119 (2022)
29. LaHaye, M.D., Suh, J., Echternach, P.M., Schwab, K.C., Roukes, M.L.: Nanomechanical measurements of a superconducting qubit. Nature **459**(7249), 960–964 (2009)
30. Moser, J., Güttinger, J., Eichler, A., Esplandiu, M.J., Liu, D.E., Dykman, M.I., Bachtold, A.: Ultrasensitive force detection with a nanotube mechanical resonator. Nat. Nanotechnol. **8**(7), 493–496 (2013)
31. De Bonis, S.L, Urgell, C., Yang, W., Samanta, C., Noury, A., Vergara-Cruz, J., Dong, Q., Jin, Y., Bachtold, A.: Ultrasensitive displacement noise measurement of carbon nanotube mechanical resonators. Nano Lett. **18**(8), 5324–5328 (2018)
32. O'Connell, A.D., Hofheinz, M., Ansmann, M., Bialczak, R.C., Lenander, M., Lucero, E., Neeley, M., Sank, D., Wang, H., Weides, M., Wenner, J., Martinis, J.M., Cleland, A.N.:.Quantum ground state and single-phonon control of a mechanical resonator. Nature **464**(7289), 697–703 (2010)
33. Palomaki, T.A., Teufel, J.D., Simmonds, R.W., Lehnert, K.W.: Entangling mechanical motion with microwave fields. Science **342**(6159), 710–713 (2013)
34. Reed, A.P., Mayer, K.H., Teufel, J.D., Burkhart, L.D., Pfaff, W., Reagor, M., Sletten, L., Ma, X., Schoelkopf, R.J., Knill, E., Lehnert, K.W.: Faithful conversion of propagating quantum information to mechanical motion. Nat. Phys. **13**(12), 1163–1167 (2017)
35. Wen, Y., Ares, N., Schupp, F.J., Pei, T., Briggs, G.A.D., Laird, E.A.: A coherent nanomechanical oscillator driven by single-electron tunnelling. Nat. Phys. **16**(1), 75–82 (2020)
36. Urgell, C., Yang, W., De Bonis, S.L., Samanta, C., Esplandiu, M.J., Dong, Q., Jin, Y., Bachtold, A.: Cooling and self-oscillation in a nanotube electromechanical resonator. Nat. Phys. **16**(1), 32–37 (2020)
37. Usmani, O., Blanter, Y.M., Nazarov, Y.V.: Strong feedback and current noise in nanoelectromechanical systems. Phys. Rev. B **75**(19), 195312 (2007)
38. Wächtler, C.W., Strasberg, P., Schaller, G.: Proposal of a realistic stochastic rotor engine based on electron shuttling. Phys. Rev. Appl. **12**(2), 024001 (2019)
39. Wächtler, C.W., Strasberg, P., Klapp, S.H.L., Schaller, G., Jarzynski, C.: Stochastic thermodynamics of self-oscillations: the electron shuttle. New J. Phys. **21**(7), 073009 (2019)
40. Wen, Y., Ares, N., Pei, T., Briggs, G.A.D., Laird, E.A.: Measuring carbon nanotube vibrations using a single-electron transistor as a fast linear amplifier. Appl. Phys. Lett. **113**(15), 153101 (2018)
41. Laird, E.A., Pei, F., Tang, W., Steele, G.A., Kouwenhoven, L.P.: A high quality factor carbon nanotube mechanical resonator at 39 ghz. Nano Lett. **12**(1), 193–197 (2012)
42. Benyamini, A., Hamo, A., Kusminskiy, S.V., von Oppen, F., Ilani, S.: Real-space tailoring of the electron–phonon coupling in ultraclean nanotube mechanical resonators. Nat. Phys. **10**(2), 151–156 (2014)
43. Wen, Y., Ares, N., Schupp, F.J., Pei, T., Briggs, G.A.D., Laird, E.A.: A coherent nanomechanical oscillator driven by single-electron tunnelling. Nat. Phys. **16**(1), 75–82 (2020)
44. Meerwaldt, H.B., Steele, G.A., van der Zant, H.S.J.: Carbon nanotubes: nonlinear high-Q resonators with strong coupling to single-electron tunneling. In: Fluctuating Nonlinear Oscillators: From Nanomechanics to Quantum Superconducting Circuits. Oxford University Press (2012)
45. Fischer, L.P., Pietzonka, P., Seifert, U.: Large deviation function for a driven underdamped particle in a periodic potential. Phys. Rev. E **97**, 022143 (2018)
46. Parrondo, J.M.R., Español, P.: Criticism of Feynman's analysis of the ratchet as an engine. Am. J. Phys. **64**(9), 1125–1130 (1996)
47. Adkins, C.J.: Equilibrium Thermodynamics. Cambridge University Press (1983)

The Role of Initial Coherence in the Phase-Space Entropy Production Rate

Giorgio Zicari, Barış Çakmak, and Mauro Paternostro

Abstract The second law of thermodynamics can be expressed in terms of entropy production, which can be used to quantify the degree of irreversibility of a process. In this Chapter, we consider the standard scenario of open quantum systems, where a system irreversibly interacts with an external environment. We show that the standard approach, based on von Neumann entropy, can be replaced by a phase-space formulation of the problem. In particular, we focus on spin systems that can be described using the so-called spin coherent states. We deploy this fomalism to study the interplay between the entropy production rate and the initial quantum coherence available to the system.

1 Introduction

First introduced by Maxwell in *The Theory of Heat*, the well-known demon named after him is the protagonist of a famous though experiment that results in a paradoxical situation where one of the pillars of classical physics—the second law of thermodynamics—is apparently violated [1]. Such a thought-provoking experiment has attracted a great deal of interest over the decades, especially given that the problem, using the language of information theory, has been rephrased in terms of information gain and feedback control [2, 3]. The picture is further complicated if one

G. Zicari (✉) · M. Paternostro
Centre for Quantum Materials and Technologies, School of Mathematics and Physics, Queen's University, Belfast BT7 1NN, UK
e-mail: G.Zicari@qub.ac.uk

B. Çakmak
Department of Physics, Farmingdale State College—SUNY, Farmingdale, New York 11735, USA

College of Engineering and Natural Sciences, Bahçeşehir University, Beşiktaş, Istanbul 34353, Turkey

M. Paternostro
Università degli Studi di Palermo, Dipartimento di Fisica e Chimica—Emilio Segrè, via Archirafi 36, 90123 Palermo, Italy

X. Bouju and C. Joachim (eds.), *Crossroad of Maxwell Demon*, Advances in Atom and Single Molecule Machines, https://doi.org/10.1007/978-3-031-57904-2_2

moves into the quantum domain, where the backaction is intimately related to the quantum measurement problem.

However, without going into the technical details of the issue, one of the questions that needs to be addressed in the fist place is the following: *which type of quantum entropy do we need to use?*

When one studies the thermodynamics of quantum systems, the standard approach is to rely on von Neumann entropy, due to its relevance for information-theoretic purposes [4]. In this Chapter, we show that an alternative approach is available, borrowing some of the well-know tools of quantum optics: the quantum phase-space formalism can be used to describe the production of entropy within a given system and between the latter and the environment that surrounds it.

In classical thermodynamics, entropy production is indeed the key quantity that allows us to express the second law of thermodynamics, providing a way to characterise and quantify the irreversibility of thermodynamic processes [5]. Let us consider the standard scenario where we deal with a macroscopic system surrounded by a thermal reservoir. Following the splitting that was put forward by Prigogine [6], one can express the infinitesimal variation in entropy dS as

$$dS = d\Sigma - d\Phi \,, \tag{1}$$

where $d\Phi$ is the entropy that the system exchanges with its surroundings, while $d\Sigma$ is the entropy intrinsically produced by the processes taking place within the system. The second law of thermodynamics imposes a constraint on the sign of $d\Sigma$, which is always non-negative (i.e. $d\Sigma \geq 0$), where the equals sign holds for reversible processes, while $d\Sigma > 0$ for irreversible transformations. Since we are interested in non-equilibrium processes, from Eq. (1) we can immediately obtain the following expression in terms of rates

$$\frac{dS}{dt} = \dot{\Sigma}(t) - \dot{\Phi}(t) \,, \tag{2}$$

where $\dot{\Sigma}(t) \geq 0$ and $\dot{\Phi}(t)$ are the *entropy production rate* and the *entropy flux rate*, respectively.

2 Entropy Production in Open Quantum Systems

2.1 From Closed to Open Systems: The Emergence of Irreversibility

Let us consider a generic quantum system S fully described by a time-independent Hamiltonian H_S. For such a system, any of its states—fully described by the density operator $\rho = \rho(t)$—evolves over time according to the following dynamical equation

$$\dot{\rho} = \mathcal{L}[\rho], \tag{3}$$

where $\mathcal{L}$ is the Liouvillian (super)-operator acting on the operator ρ, which we assume to have been prepared in a given initial state $\rho_0 = \rho(t = 0)$.

In the ideal case where the system is completely isolated, Eq. (3) assumes the form of a Liouville–von Neumann equation with

$$\mathcal{L}[\rho] \equiv -i[H_S, \rho]. \tag{4}$$

In this scenario, the system's dynamics is fully reversible: the equations of motion are invariant under the time-reversal symmetry transformation $t \to -t$. This can be seen straightforwardly by considering a two-level system (i.e. a qubit) with energy described by the Hamiltonian $H_S = (\omega_0/2)\sigma_x$.[1] Here ω_0 is the energy difference between the two levels, while σ_j is the $j = x, y, z$ Pauli operator. The system, initially prepared in one of the eigenstates of σ_z, i.e. either $|0\rangle$ or $|1\rangle$, oscillates indefinitely between such states—cf. Fig. 1.

Now we let our system S—which we assume to be fully within our control—interact with a finite (small) number N of otherwise unaccessible quantum harmonic oscillators (hereafter dubbed *modes*), initially prepared in a thermal state at a given temperature T. We are thus setting a boundary between S and a second subsystem B, which can be called *environment*, made out of the collection of modes. This has a precise physical meaning, as it implies focusing onto S (the so-called *reduced* system) while effectively discarding the environmental degrees of freedom. Mathematically, it is tantamount to performing a partial trace using the basis of the Hilbert space associated with B. Therefore, by explicitly solving Eq. (4) for the state $\rho_{SB}(t)$ of the $S - B$ compound, we can obtain $\rho_S(t) = \mathrm{tr}_B\, \rho_{SB}(t)$ at any time t. By doing so, we depart from the totally reversible scenario we previously had when S was perfectly isolated—this becomes more and more evident as the number of modes N increases. The truly irreversible dynamics is obtained when we take the limit $N \to \infty$.

In principle, the operation of tracing out the environment can be done in a (numerically) exact way. However, this becomes quickly unfeasible as the number of modes increases. Luckily enough, the theory of open quantum systems provides us with an effective way to perform this task. By making some further assumptions—collectively known as *Born-Markov approximation* [7]—we are able to track the evolution of the reduced system. We should stress that this comes at a cost: we are coarse-graining the system dynamics assuming that the environmental correlation functions decay much faster than the typical timescale over which our system evolves. Physically, it means that the information is monotonically flowing from the system to the environment, leading to a Markovian (or memoryless) dynamical process [7, 8]. Such a separation of the timescales allows us to derive special classes of dynamical equations, known as master equations, that can be still brought in the form of Eq. (3), where the Liouvillian reads as

[1] We assume units such that $\hbar = k_B = 1$.

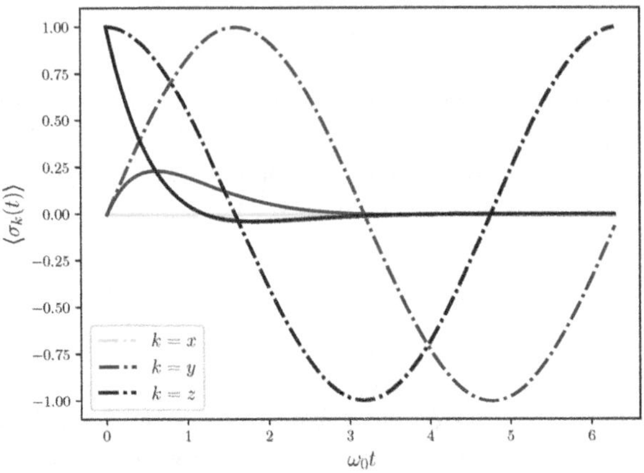

Fig. 1 Time evolution of the relevant observables $\langle\sigma_k(t)\rangle$ (with $k = x, y, z$) of a qubit. The solid lines refer to the closed dynamics governed by Eq. (4), while dotted-dashed lines refer to the dissipative dynamics, as described by a dissipator of the form of Eq. (6). The plots clearly shows that the interaction with the thermal bath leads to irreversibility, causing the observables to depart from the periodic oscillations of the prefectly isolated scenario

$$\mathcal{L}[\rho] \equiv -i[H_S, \rho] + D(\rho)\,, \tag{5}$$

with the first term accounting for the unitary dynamics, while $D(\rho)$ being the so-called dissipator, which effectively includes the environmental effects, such as decoherence and/or dissipation. In particular, we would like to focus on dissipators of a special form, known as *Lindblad-Davies maps*, which play a important role in quantum thermodynamics. They can be expressed as

$$D(\rho) = \sum_j \Gamma_j^- \left(L_j^- \rho L_j^+ - \frac{1}{2}\{L_j^+ L_j^-, \rho\} \right) + \sum_j \Gamma_j^+ \left(L_j^+ \rho L_j^- - \frac{1}{2}\{L_j^- L_j^+, \rho\} \right), \tag{6}$$

where $L_j^{\pm}$ are the Lindblad operators describing the coupling between the system S and the various decay channels with damping rates $\Gamma_j^{\pm}$ [9]. The latter are assumed to satisfy the local detailed balance condition

$$\frac{\Gamma_j^+}{\Gamma_j^-} = e^{-\beta\omega_j} \tag{7}$$

for each j, where β is the inverse temperature of the thermal reservoir. Once again we can consider the example of a qubit, where we consider only two decay channels, with rates $\Gamma^- = \Gamma(\bar{n} + 1)$ and $\Gamma^+ = \Gamma\bar{n}$, accounting for the incoherent loss and gain of excitations, respectively. Note that $\bar{n} = (e^{\beta\omega} - 1)^{-1}$ is the average number of excitations in the bath. The corresponding Lindblad operators are given by $L^{\pm} = \sigma_{\pm} = (\sigma_x \pm i\sigma_y)/2$. The time evolution of the observables clearly shows the irreversible character of the dynamics, as shown in Fig. 1.

2.2 *Standard Approach: Von Neumann Entropy Production*

When we move from an isolated to an open system, we allow the main system to exchange energy—or, generally speaking, *information*—with its surroundings. To quantify the entropy production rate and flux, we need to introduce a suitable definition of entropy [10]. The standard choice for quantum information processes is given by the von Neumann entropy, which—for a state ρ—is defined as

$$S_{\mathrm{vN}}(\rho) = -\operatorname{Tr}(\rho \ln \rho). \tag{8}$$

By taking the time derivative of the latter, we readily obtain $\dot{S}_{\mathrm{vN}}(\rho) = -\operatorname{Tr}(\dot{\rho} \ln \rho)$. Since $\dot{\rho} = \mathcal{L}[\rho]$, we obtain

$$\dot{S}_{\mathrm{vN}}(\rho) = -\operatorname{Tr}(\mathcal{L}[\rho] \ln \rho)\,. \tag{9}$$

The entropy flux rate is associated with the heat flow, which, in the weak coupling limit, is defined as $\dot{Q} \equiv -\operatorname{Tr}(H_S \dot{\rho})$. If the system is weakly coupled to a thermal reservoir at inverse temperature β, the system relaxes towards the local Gibbs state given by $\rho_{\mathrm{eq}} = e^{-\beta H_S}/\mathcal{Z}_\beta$, where $\mathcal{Z}_\beta \equiv \operatorname{Tr}(e^{-\beta H_S})$ is the partition function.

As in the classical case, we have

$$\dot{\Phi}(t) = \beta\, \dot{Q}(t) = \operatorname{Tr}\left(\mathcal{L}[\rho] \ln \rho_{\mathrm{eq}}\right), \tag{10}$$

where we have resorted to the identity $\ln \rho_{\mathrm{eq}} = -\beta H_S - \mathbb{1} \ln \mathcal{Z}_\beta$ and the fact that $\operatorname{Tr} \dot{\rho} = 0$.

Equations (9)–(10) lead to the following expression for the entropy production rate:

$$\dot{\Sigma}(t) = \dot{S}_{\mathrm{vN}}(\rho) + \dot{\Phi}(t) = -\operatorname{Tr}\left[\mathcal{L}[\rho]\left(\ln \rho - \ln \rho_{\mathrm{eq}}\right)\right] \tag{11}$$

which can be recast in the form of a time derivative, i.e.

$$\dot{\Sigma}(t) = -\frac{d}{dt} S(\rho || \rho_{\mathrm{eq}})\,, \tag{12}$$

provided that we introduce the von Neumann relative entropy, defined as $S_{\mathrm{vN}}(\varrho||\varsigma) \equiv \operatorname{Tr}\{\varrho(\ln \varrho - \ln \varsigma)\}$. For a quantum system described by a Liouvillian in the form of Eq. (5), from the so-called Spohn inequality it follows that $\dot{\Sigma}(t) \geq 0$ at all times, as required by the second law of thermodynamics [10]. From Eqs. (10) and (11), it is immediate to conclude that $\dot{\Phi}(t) = 0 = \dot{\Sigma}(t)$ when the system reaches the local Gibbs state, as $\mathcal{L}[\rho_{\mathrm{eq}}] = 0$.

2.3 The Role Played by Quantum Coherence

It can be explicitly shown that a master equation of the form specified by Eq. (5) can be derived *ab initio* starting from the microscopic Hamiltonian describing the system, its environment, and the interaction between them. By working in the Born-Markov approximation, one obtains two separate sets of dynamical equations: one governing the diagonal entries of ρ (i.e. the *populations*), the other governing the off-diagonal entries (i.e. the *coherences*) [7]. This mathematical evidence expresses the occurrence of two different physical processes dynamically taking place: on one hand, there are transitions between the energy levels, causing the populations to adjust to values imposed by the thermal bath; on the other hand, one witnesses the loss of coherence in the energy eigenbasis.

For the sake of definiteness, let us consider the eigenbasis $\{|n\rangle\}$. To be more quantitative, the populations $P_n \equiv \langle n|\rho|n\rangle$ behave classically, obeying the Pauli master equation [11]

$$\frac{dP_n}{dt} = \sum_n \left[W(n|k)P_k(t) - W(k|n)P_n(t)\right] , \tag{13}$$

where $W(n|k)$ are the transition rates from the energy level ϵ_n to the level ϵ_k, satisfying the detailed balance condition

$$\frac{W(n|k)}{W(k|n)} = \frac{P_n^{\text{eq}}}{P_k^{\text{eq}}} = e^{-\beta(\epsilon_n - \epsilon_k)} , \tag{14}$$

where $P_n^{\text{eq}} \equiv \langle n|\rho_{\text{eq}}|n\rangle$.

Inspired by the general definition of free energy, i.e. $F = E - TS$, we can define the following non-equilibrium free energy [12]

$$F(\rho) = \text{Tr}\,(H_S\rho) + T\,\text{Tr}\,(\rho \ln \rho) \tag{15}$$

where T is the temperature of the bath. It is immediate to check that, for equilibrium states, one recovers the well known result of statistical mechanics

$$F_{\text{eq}} \equiv F(\rho_{\text{eq}}) = -T \ln \mathcal{Z}_\beta , \tag{16}$$

whose generalisation for non-equilibrium states reads as

$$F(\rho) = F_{\text{eq}} + TS(\rho||\rho_{\text{eq}}) . \tag{17}$$

Since $S(\rho||\rho_{\text{eq}}) \geq 0$, we readily obtain the condition $F(\rho) \geq F_{\text{eq}}$, which automatically defines the equilibrium state ρ_{eq} as the one minimising the free energy.

Combining Eqs. (12) and (17), we obtain the following expression for the entropy production rate

$$\dot{\Sigma}(t) = -\frac{1}{T}\frac{dF(\rho)}{dt}. \tag{18}$$

Note that the state ρ_{eq} is diagonal in the eigenbasis $\{|n\rangle\}$ of H_S; differently, the state ρ is in general an incoherent state which can be decomposed into a diagonal part ρ_{diag} and a off-diagonal one χ, i.e. $\rho = \rho_{\text{diag}} + \chi$. Therefore, we can introduce the relative entropy of coherence [13], defined as

$$C(\rho) = S(\rho_{\text{diag}}) - S(\rho), \tag{19}$$

which allows us to split the relative entropy as

$$S(\rho||\rho_{\text{eq}}) = S(\rho_{\text{diag}}||\rho_{\text{eq}}) + C(\rho), \tag{20}$$

where $S(\rho_{\text{diag}}||\rho_{\text{eq}})$ is the Kullback-Leibler divergence of the classical probability distribution $\{P_n\}_n$ relative to the equilibrium distribution $\{P_n^{\text{eq}}\}_n$, i.e.

$$S(\rho_{\text{diag}}||\rho_{\text{eq}}) = \sum_n P_n \ln\left(\frac{P_n}{P_n^{\text{eq}}}\right). \tag{21}$$

Therefore, Eq. (17) can be rewritten as

$$F(\rho) = F_{\text{eq}} + TS(\rho_{\text{diag}}||\rho_{\text{eq}}) + TC(\rho), \tag{22}$$

which shows that, apart from the equilibrium contribution F_{eq}, the non-equilibrium free energy is made of two separate contributions: one classical and one genuinely quantum, corresponding to the second and third term of Eq. (22), respectively. The former quantifies the increase in free energy due to the population imbalance with respect to the equilibrium configuration; the latter, instead, expresses the additional free energy contribution carried by a state with non-null coherences [12].

It is straightforward to show that the splitting introduced by Eq. (22) carries over to the entropy production rate

$$\dot{\Sigma}(t) = -\frac{d}{dt}S(\rho_{\text{diag}}||\rho_{\text{eq}}) - \frac{dC(\rho)}{dt} \equiv \dot{\Sigma}_d(t) + \Upsilon(t), \tag{23}$$

where, by resorting to Eq. (21), we obtain the following expression for the classical contribution to the entropy production rate

$$\dot{\Sigma}_d(t) = \frac{1}{2}\sum_{n,k}[W(n|k)P_k - W(k|n)P_n]\ln\left(\frac{W(k|n)P_n}{W(n|k)P_k}\right), \tag{24}$$

where we have used the Pauli master equation (13) together with the detailed balance condition (14). The contribution due to quantum coherence is simply given by $\Upsilon(t) = -dC(t)/dt$.

3 Entropy Production in the Quantum Phase Space

The standard approach to entropy production in open quantum systems described in Sect. 2 provides a consistent framework based on von Neumann entropy. We should stress that the aforementioned results are based on a certain number of identifications and definitions which are valid if and only if we deal with quantum systems weakly coupled with a thermal reservoir, as far as we satisfy all the assumptions of the Born-Markov approximation. However, even when we move within that approximation, the framework we have just introduced is not flawless. For instance, in the limit of zero temperature reservoirs (i.e. $\beta \to \infty$) this framework is formally inconsistent. On one hand, the reference state in the relative entropy, namely ρ_{eq}, becomes pure, which causes $S(\rho||\rho_{\mathrm{eq}})$ and $\dot{\Sigma}(t)$ [cf. Eq. (12)] to be ill-defined. The entropy flux rate $\dot{\Phi}(t)$ is also divergent, as one can easily conclude from Eq. (10). However, the overall rate dS/dt stays finite, showing that the aforementioned divergences have no corresponding physical meaning: the zero temperature limit is indeed frequently considered in quantum optics [14]. A similar inconsistency is found whenever the system is in a pure state.

Therefore, one possible solution to this problem rely on a semi-classical formulation for the entropy production in the quantum phase-space, which coincides with the standard approach at high temperatures. The core idea is to replace the von Neumann entropy with a generalised entropy function, associated with a certain probability distribution defined over the phase space [15, 16]. In particular, hereafter we will refer to the Wehrl entropy defined using the Husimi-Q function [17]. We will resort to the latter to describe spin systems for which a phase-space formulation is available, namely those described through the so-called spin coherent states.

3.1 Spin Coherent States

Let us consider a single quantum system described by the operators J_x, J_y, J_z satisfying the algebra $[J_x, J_y] = i J_z$. A *spin coherent state* is defined as [18]

$$|\Omega\rangle = e^{-i\phi J_z} e^{-i\theta J_y} |J\rangle , \qquad (25)$$

where $\Omega = (\theta, \phi)$ is a solid angle in polar coordinates (with $\theta \in [0, \pi]$ and $\phi \in [0, 2\pi]$), while $|J\rangle$ is the angular momentum state with the largest quantum number of J_z.

If the system is described by the density operator ρ, we can define the Husimi Q-function $\mathcal{Q} = \mathcal{Q}(\Omega)$ over the phase space

$$\mathcal{Q}(\Omega) \equiv \langle\Omega|\rho|\Omega\rangle. \qquad (26)$$

Furthermore, if the system dynamics is governed by a master equation in the Lindblad form, as in Eq. (5), the phase-space counterpart of the latter assumes the form of a Fokker-Planck equation, i.e.

$$\partial_t \mathcal{Q} = \mathcal{U}(\mathcal{Q}) + \mathcal{D}(\mathcal{Q})\,, \tag{27}$$

where $\mathcal{U}(\mathcal{Q})$ and $\mathcal{D}(\mathcal{Q})$ account for the unitary and dissipative parts of the evolution, respectively.

If the system evolves according to a standard open system dynamics described by the generic Liouvillian in Eq. (5), one can derive a Fokker-Planck equation of the form (27) using a set of suitable correspondence rules. For instance, we have [18]

$$\begin{aligned}
\left[J_+, \rho\right] &\rightarrow \mathcal{J}_+(\mathcal{Q}) = e^{i\phi}\left(\partial_\theta + i\cot\theta\, \partial_\phi\right)\mathcal{Q}\,, \\
\left[J_-, \rho\right] &\rightarrow \mathcal{J}_-(\mathcal{Q}) = -e^{-i\phi}\left(\partial_\theta - i\cot\theta\, \partial_\phi\right)\mathcal{Q}\,, \\
\left[J_z, \rho\right] &\rightarrow \mathcal{J}_z(\mathcal{Q}) = -i\partial_\phi \mathcal{Q}\,,
\end{aligned} \tag{28}$$

where $J_\pm = J_x \pm i J_y$.

3.2 *Spin-Phase-Space Entropy Production*

Given the Husimi Q-function $\mathcal{Q} = \mathcal{Q}(\Omega)$, one can define the Wehrl entropy as [17]

$$\mathcal{S}_\mathcal{Q} = -\left(\frac{2J+1}{4\pi}\right)\int d\Omega\, \mathcal{Q} \ln \mathcal{Q}\,, \tag{29}$$

where the prefactor is chosen just for convenience. By taking the time derivative of Eq. (29), the normalisation condition together with Eq. (27) yields

$$\left.\frac{d\mathcal{S}_\mathcal{Q}}{dt}\right|_{\text{diss}} = -\left(\frac{2J+1}{4\pi}\right)\int d\Omega\, \mathcal{D}(\mathcal{Q}) \ln \mathcal{Q}\,, \tag{30}$$

where we consider only the contribution coming from the dissipative part [16]. For a given dynamical process, the idea is to bring Eq. (30) into the Prigogine form of Eq. (2), separating the entropy production rate $\dot{\Sigma}(t)$ from the flux rate $\dot{\Phi}(t)$.

4 Spin-Phase-Space Entropy Production Rate

In this Section, we derive the explicit expressions for the entropy production rate for two relevant classes of open system dynamics: dephasing and amplitude damping channels [4]. In order to do that, we start from Eq. (30) where we include only the contribution coming from the dissipative part of the dynamics, then we identify $\dot{\Sigma}$ and $\dot{\Phi}$ according to the splitting introduced in Eq. (2).

4.1 Dephasing Channels

Let us consider the case of dephasing channels, where the dissipator reads as

$$D(\rho) = -\frac{\lambda}{2}\left[J_z, \left[J_z, \rho\right]\right], \tag{31}$$

with λ being the dephasing rate. Note that a dissipator of this form does not entail any direct energy exchange between the system and the bath: the interaction between them causes the system to loose coherence. Using the correspondence rules (28), one can obtain the following phase-space dissipator

$$\mathcal{D}(\mathcal{Q}) = -\frac{\lambda}{2}\mathcal{J}_z\left(\mathcal{J}_z(\mathcal{Q})\right). \tag{32}$$

By plugging Eq. (32) into Eq. (30), we get

$$\dot{\Sigma} \equiv \left.\frac{dS_{\mathcal{Q}}}{dt}\right|_{\text{diss}} = \frac{\lambda}{2}\left(\frac{2J+1}{4\pi}\right)\int \frac{|\mathcal{J}_z(\mathcal{Q})|^2}{\mathcal{Q}}d\Omega, \tag{33}$$

after having performed an integration by parts. From Eq. (33), it is immediate to conclude that $\dot{\Sigma} \geq 0$, as requested from the second law of thermodynamics. Note that $\dot{\Sigma} = 0$ if and only if $\mathcal{J}_z(\mathcal{Q}) = 0$, which occurs when the function $\mathcal{Q} = \mathcal{Q}(\theta, \phi)$ is independent of the azimuthal angle ϕ, as, by definition, we have $\mathcal{J}_z(\mathcal{Q}) = -i\partial_\phi\mathcal{Q}$. In the phase space, the latter is equivalent to the condition of ρ being diagonal in the J_z basis. Therefore, $\mathcal{J}_z(\mathcal{Q})$ is the phase-space current associated to the genuine quantum phenomenon of decoherence. Moreover, it should be stressed that, for this channel, the identification of $\dot{\Sigma}$ is straightforward, as the entropy flux rate $\dot{\Phi}$ is identically zero, due to the lack of energy exchange between the system and the bath.

4.2 Amplitude Damping Channels

We consider a second relevant example, i.e. amplitude damping channels, where the physical picture is richer: during the dynamical process, the populations adjust to values imposed by the thermal bath, while we witness the incoherent exchange of thermal excitations between the system and the environment. Such a process is mathematically expressed by a dissipator of the form

$$D(\rho) = \Gamma(\bar{n}+1)\left(J_-\rho J_+ - \frac{1}{2}\{J_+J_-, \rho\}\right) + \Gamma\bar{n}\left(J_+\rho J_- - \frac{1}{2}\{J_-J_+, \rho\}\right), \tag{34}$$

where Γ is the damping rate, while $\bar{n} = (e^{\beta\omega_0} - 1)^{-1}$ is the average number of thermal excitations in the bath. Note that this dissipator is a prototypical example of a Lindblad-Davis map, as in Eq. (6). A more involved derivation (which can be found in Ref. [16]) shows that the phase-space dissipator reads as

$$\mathcal{D}(\mathcal{Q}) = \frac{\Gamma}{2}\left\{\mathcal{J}_-(f(\mathcal{Q})) - \mathcal{J}_+(f^*(\mathcal{Q}))\right\}, \tag{35}$$

where the phase-phase operators $\mathcal{J}_\pm$ are defined in Eq. (28), while

$$f(\mathcal{Q}) = \frac{1}{2}\left[2\,J\mathcal{Q} - \mathcal{J}_z(\mathcal{Q})\right]e^{i\phi}\sin\theta + \frac{1}{2}\left[\cos\theta - (2\bar{n}+1)\right]\mathcal{J}_+(\mathcal{Q}). \tag{36}$$

By substituting Eq. (35) into Eq. (30) and integrating by parts, we obtain

$$\left.\frac{dS_{\mathcal{Q}}}{dt}\right|_{\text{diss}} = \frac{\Gamma}{2}\left(\frac{2J+1}{4\pi}\right)\int\frac{\mathcal{F}(\mathcal{Q})}{\mathcal{Q}}d\Omega, \tag{37}$$

where $\mathcal{F}(\mathcal{Q}) = f(\mathcal{Q})\mathcal{J}_-(\mathcal{Q}) - f^*(\mathcal{Q})\mathcal{J}_+(\mathcal{Q})$.

We can then separate $\dot{\Sigma}$ from $\dot{\Phi}$ as in Eq. (2). The rationale behind this choice is that the entropy production and flux rates should be even and odd functions of the relevant currents, respectively, as expected from standard non-equilibrium thermodynamics arguments [6]. The calculations, detailed in Ref. [16], lead to the following expression for the entropy production rate

$$\dot{\Sigma} = \frac{\Gamma}{2}\left(\frac{2J+1}{4\pi}\right)\int\frac{d\Omega}{\mathcal{Q}}\left\{\frac{\{2J\mathcal{Q}\sin\theta + [\cos\theta - (2\bar{n}+1)]\partial_\theta\mathcal{Q}\}^2}{(2\bar{n}+1)-\cos\theta} + |\mathcal{J}_z(\mathcal{Q})|^2\left[(2\bar{n}+1)\cos\theta - 1\right]\frac{\cos\theta}{\sin^2\theta}\right\}. \tag{38}$$

This rather cumbersome expression for the entropy production rate is process-specific, in the sense that it applies to the specific type of dynamics we have considered. As a consequence, the mathematical expression contained in Eq. (38) provides some insight on the processes taking place during the time evolution. It is indeed clear that there are two distinct contributions: one is proportional to the dephasing current $|\mathcal{J}_z(\mathcal{Q})|^2$, which accounts for the loss of coherence; the other directly related to the amplitude damping.

5 Influence of the Initial Coherence on the Entropy Production Rate

In this Section, we aim to study the influence of the initial coherence on the spin-phase-space entropy production rate. To this end, we will consider the expressions for the spin-phase entropy production rate derived in Sect. 4. For a given dynamical process (either dephasing or amplitude damping channels), we initially prepare the system in a state $\rho_0 = \rho_{\text{diag}} + \chi$, where we explicitly separate the populations ρ_{diag} (i.e. the classical part) from the coherences χ (i.e. the quantum part). To quantify the initial coherence available to the system we use the l_1-norm [13], i.e.

$$\mathcal{C}(\rho_0) = \sum_{i,j} |\chi_{ij}| \, . \tag{39}$$

We then solve the dynamical equations $\dot{\rho} = \mathcal{L}[\rho]$, whence we obtain the time evolution of the entropy production rate, i.e., $\dot{\Sigma} = \dot{\Sigma}(t)$.

5.1 Qubit

Let us consider the simplest case of a single qubit (i.e. a single spin$-1/2$ particle), for which the most general density matrix can be written in the form

$$\rho = \frac{1}{2} \left(\mathbb{1} + \boldsymbol{\sigma} \cdot \boldsymbol{\tau} \right) , \tag{40}$$

where $\boldsymbol{\sigma} = (\sigma_x, \sigma_y, \sigma_z)$ and $\boldsymbol{\tau} = (\tau_x, \tau_y, \tau_z)$ are the Pauli and Bloch vectors, respectively, where $\tau_i = \text{Tr}(\sigma_i \rho)$. We can construct the spin coherent state $|\Omega\rangle$ using the definition in Eq. (25), where we take $J_i = \sigma_i/2$, and $|J\rangle = |0\rangle$ is the eigenstate of σ_z corresponding to the eigenvector $+1$. From Eq. (26), the Husimi Q-function reads as

$$\mathcal{Q}(\Omega) = \frac{1}{2} \left(\mathbb{1} + \hat{n} \cdot \boldsymbol{\tau} \right) , \tag{41}$$

where $\hat{n} = (\sin\theta\cos\phi, \sin\theta\sin\phi, \cos\theta)$ is the unit vector. Using Eq. (39), the coherence available to the initial state can be immediately quantified as $\mathcal{C}(\rho_0) = 2(\tau_x^2 + \tau_y^2)$, where τ_x and τ_y are the first two Bloch components of the density matrix ρ_0 at time $t = 0$.

Let us consider the case of purely dephasing dynamics expressed by the dissipator in Eq. (31), where $J_z = \sigma_z/2$. By plugging Eq. (41) into Eq. (33), a cumbersome integration over the phase-space yields [16]

$$\dot{\Sigma} = \frac{\lambda}{4} (\tau_x^2 + \tau_y^2) \left\{ \frac{\tau - (1 - \tau^2) \tanh^{-1}(\tau)}{\tau^3} \right\} , \tag{42}$$

where $\tau = (\tau_x^2 + \tau_y^2 + \tau_z^2)^{1/2}$. By contrast, the von Neumann entropy production is given by [16]

$$\dot{\Sigma}_{\rm vN} = \frac{\lambda}{2}(\tau_x^2 + \tau_y^2)\frac{\tanh^{-1}(\tau)}{\tau}\,. \tag{43}$$

Note that, if the state of the system is pure, i.e. $\tau \to 1$, while the von Neumann entropy production diverges, the phase-space counterpart stays finite, namely $\dot{\Sigma} \to \lambda \sin^2\theta/4$, consistently with what stated at the beginning of Sect. 4.

Analogously, one can derive the closed expression of $\dot{\Sigma}$ for the amplitude damping channels, by considering Eq. (34) with $J_\pm = \sigma_\pm$. By plugging Eq. (41) into Eq. (38), after integration, we eventually get [16]

$$\dot{\Sigma} = \frac{\Gamma}{2}\frac{2\bar{\tau}_z\tau_z - (\tau^2 + \tau_z^2)}{2\bar{\tau}_z}\left[\frac{\tau - (1-\tau^2)\tanh^{-1}(\tau)}{\tau^3}\right] + \frac{\Gamma}{2}(\tau_z - \bar{\tau}_z)\left[\frac{\bar{\tau}_z - (1-\bar{\tau}_z^2)\tanh^{-1}(\bar{\tau}_z)}{\bar{\tau}_z^3}\right], \tag{44}$$

where we have introduced the bath-induced magnetisation $\bar{\tau}_z \equiv -1/(2\bar{n}+1)$. Differently, one can show that the von Neumann entropy production rate reads as [16]

$$\dot{\Sigma}_{\rm vN} = \Gamma\frac{\tanh^{-1}(\bar{\tau}_z)}{\bar{\tau}_z}(\tau_z - \bar{\tau}_z) - \frac{\Gamma}{2}\frac{\tanh^{-1}(\tau)}{\tau\bar{\tau}_z}\left[\tau^2 + \tau_z(\tau_z - 2\bar{\tau}_z)\right]. \tag{45}$$

For both classes of channels, we can plot the entropy production rates $\dot{\Sigma}$ and $\dot{\Sigma}_{\rm vN}$ as functions of the initial coherence $C(\rho_0)$. Figure 2 highlights the monotonic relationship between the quantum coherence we input and the entropy production rate.

One can also look at the dynamics of the entropy production rate. For the dephasing channel, the explicit solution of the dynamics in the interaction picture yields

$$\tau_i(t) = \tau_i(0)e^{-\lambda t/2} \quad i = x, y\,, \tag{46}$$

while the third component stays constant over time, i.e. $\tau_z(t) = \tau_z(0)$. By plugging $\boldsymbol{\tau}(t) = (\tau_x(t), \tau_y(t), \tau_z(t))$ back into Eq. (42), one can track the evolution of the phase-space entropy production rate. In Fig. 3a, we prepare the system in different initial states, labelled by a different value of the intial coherence $C(\rho_0)$: one can immediately conclude that the higher is the initial coherence, the higher the entropy production rate.

Similarly, one can solve the amplitude damping dynamics, where all the components of the Bloch vector evolve non-trivially to eventually reach thermalisation. Explicitly, one gets

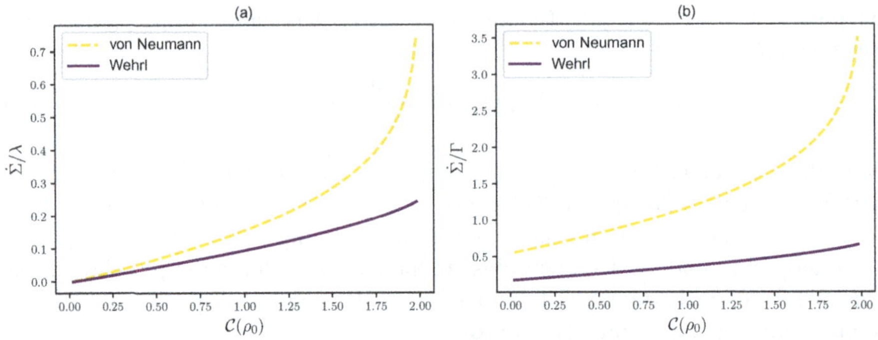

Fig. 2 Entropy production rate $\dot{\Sigma}$ as a function of the coherence available to the system, as measured through the l_1-norm quantifier, i.e. $\mathcal{C}(\rho_0) = 2(\tau_x^2 + \tau_y^2)$. In Panel (**a**), we consider the case of a system undergoing pure dephasing dynamics, as described by the dissipator in Eq. (31). In Panel (**b**), we consider the case of amplitude damping channels described by Eq. (34) instead. In both cases, we compare the von Neumann entropy production rate and its Wehrl counterpart. Note that the latter underestimates the entropy production rate, while the former diverges whenever we approach the limit of a pure state. For the plots displayed in figure, we chose $\tau_z = 0$ and $\bar{\tau}_z = 0$

$$\tau_i(t) = \tau_i(0)e^{-\frac{\Gamma}{2}(2\bar{n}+1)t} \quad i = x, y \,, \tag{47}$$

$$\tau_z(t) = \tau_z(0)e^{-\Gamma(2\bar{n}+1)t} + \bar{\tau}_z \,. \tag{48}$$

If we plug these equations into Eqs. (44), we obtain $\dot{\Sigma} = \dot{\Sigma}(t)$. Similarly to the case of pure dephasing channels, we consider different initial states, namely, different τ_x and τ_y, while we keep τ_z fixed, as the latter enters in the diagonal part of the density matrix, which behaves classically. This case shows a similar behaviour: there is a monotonicity relationship between the coherence we input and the entropy production rate.

5.2 *Qutrit*

Furthermore, we can consider the case of a qutrit, i.e. a spin-1 particle. In this case, for the sake of simplicity, we can assume that all the entries of the density matrix are real numbers, i.e.

$$\rho_0 = \begin{pmatrix} \rho_{00} & \alpha & \beta \\ \alpha & \rho_{11} & \gamma \\ \beta & \gamma & \rho_{22} \end{pmatrix} , \tag{49}$$

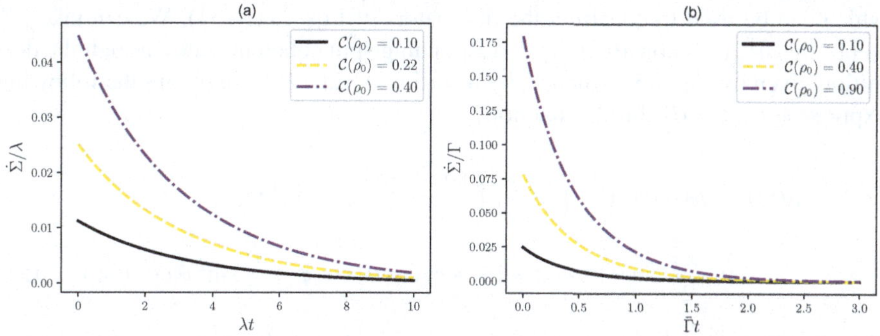

Fig. 3 Wehrl entropy production rate $\dot{\Sigma}$ as a function of time. In both Panels, we show several curves corresponding to different values of the initial coherence, as quantified by the l_1-norm, i.e. $\mathcal{C}(\rho_0) = 2(\tau_x^2 + \tau_y^2)$. In Panel (**a**), we consider a single qubit subject to pure dephasing, while in Panel (b) we consider we the case of amplitude damping dynamics. For the plots shown in Panel (a), we took different values of τ_x and τ_y to reproduce the values of coherence displayed in the legend, while τ_z has been taken so that $\tau^2 = 0.9$. In Panel (**b**), we chose τ_x and τ_y with the same rationale, while $\tau_z = 0.1$. The average number of excitations in the bath is given by $\bar{n} = 0.5$, while $\bar{\Gamma} = \Gamma(2\bar{n} + 1)$

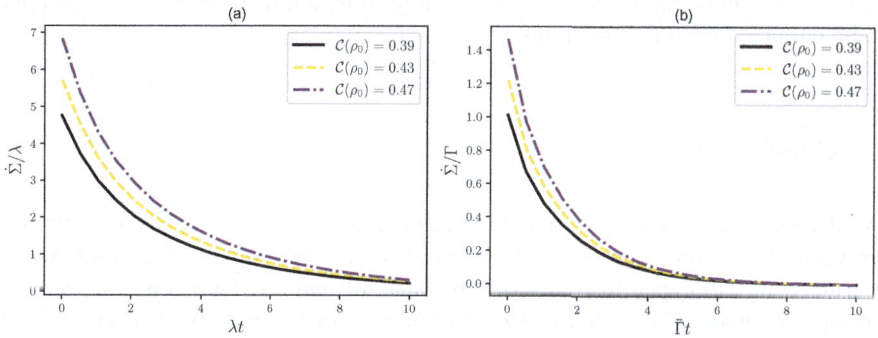

Fig. 4 Wehrl entropy production rate $\dot{\Sigma}$ as a function of time for a qutrit. In both Panels, we show several curves corresponding to different values of the initial coherence, as quantified by the l_1-norm, i.e. $\mathcal{C}(\rho_0) = 2(|\alpha| + |\beta| + |\gamma|)$. In Panel (**a**), we consider a single qutrit subject to pure dephasing, while in Panel (**b**) we consider we the case of amplitude damping dynamics. The initial states are randomly generated so that we reproduce the value of coherence shown in the legend. The average number of excitations in the bath is given by $\bar{n} = 0.5$, while $\bar{\Gamma} = \Gamma(2\bar{n} + 1)$

thus, according to the l_1-norm, the quantum coherence associated with it is given by $\mathcal{C}(\rho_0) = 2(|\alpha| + |\beta| + |\gamma|)$. In this case, we can represent the spin operators as

$$J_x = \frac{1}{\sqrt{2}}\begin{pmatrix} 0 & 1 & 0 \\ 1 & 0 & 1 \\ 0 & 1 & 0 \end{pmatrix}, \quad J_y = \frac{1}{\sqrt{2}}\begin{pmatrix} 0 & -i & 0 \\ i & 0 & -i \\ 0 & i & 0 \end{pmatrix}, \quad J_z = \begin{pmatrix} 1 & 0 & 0 \\ 0 & 0 & 0 \\ 0 & 0 & -1 \end{pmatrix}, \tag{50}$$

which can be used to construct the dissipators of Eqs. (31)–(34). We can take $|J\rangle$ as $(1 \quad 0 \quad 0)^{\mathrm{T}}$, then construct the corresponding spin coherent state through the definition given by Eq. (25), whence, by resorting to Eq. (26), we obtain the following expression for the Husimi Q-function

$$\mathcal{Q}(\Omega) = \rho_{00} \cos^4\left(\frac{\theta}{2}\right) + \rho_{11}\left(\frac{\sin^2\theta}{2}\right) + \rho_{22} \sin^4\left(\frac{\theta}{2}\right) \\ \times \frac{\sqrt{2}}{2} \sin\theta\,[a + c + (a - c)\cos\theta]\cos\phi + \frac{b}{2}\sin^2\theta\cos(2\phi). \quad (51)$$

This function, together with its derivatives, can be used to calculate the entropy production rate for the cases of dephasing and amplitude damping channels, through Eqs. (33)–(38). Similarly to the case of a single qubit, one can track the entropy production rate over time, i.e. $\dot{\Sigma} = \dot{\Sigma}(t)$. One needs to explicitly solve the dynamics and calculate the corresponding entropy production rate at each time step by using Eqs. (33)–(38), and integrating over the phase space. For both the dephasing and the amplitude damping channels, one can prepare the system into different initial states (characterised by a different amount of coherence) and determine $\dot{\Sigma} = \dot{\Sigma}(t)$. In Fig. 4 we show that initial states characterised by a larger coherence correspond to higher entropy production rates.

6 Conclusions

We discussed how entropy production can be used to characterise and quantify the irreversibility arising when we consider a quantum system weakly interacting with a thermal bath. More specifically, beside the standard approach based on von Neumann entropy, we resorted to a phase-space formulation of the entropy production. Such description proved useful to attack the case of spin systems that can be described by means of the so-called spin coherent states. We proved that—both for dephasing and amplitude damping channels—the initial coherence available to the system is monotonically related to the entropy production rate. However, such a relationship is not general. For instance, as shown in Ref. [19], for randomly generated initial states of bipartite spin systems, there might be instances contradicting the direct proportionality between initial coherence and high entropy production rates.

Acknowledgements BÇ is partially supported by The Scientific and Technological Research Council of Turkey (TUBITAK) under Grant No. 121F246. MP acknowledges the support by the European Union's Horizon 2020 FETOpen project TEQ (766900), the Horizon Europe EIC Pathfinder project QuCoM (Grant Agreement No. 101046973), the Leverhulme Trust Research Project Grant UltraQuTe (grant RGP-2018-266), the Royal Society Wolfson Fellowship (RSWF/R3/183013), the UK EPSRC (EP/T028424/1), and the Department for the Economy Northern Ireland under the US-Ireland R&D Partnership Programme.

References

1. Maxwell, J.C.: The Theory of Heat. Longmans, Green, New York (1988)
2. Landauer, R.: Irreversibility and heat generation in the computing process: IBM. J. Res. Dev. **5**, 183 (1961)
3. Bennett, C.H.: Logical reversibility of computation: IBM. J. Res. Dev. **17**, 525 (1973)
4. Nielsen, M.A., Chuang, I.L.: Quantum Computation and Quantum Information: 10th Anniversary Edition. Cambridge University Press, Cambridge (2010)
5. Landi, G.T., Paternostro, M.: Irreversible entropy production: from classical to quantum. Rev. Mod. Phys. **93**, 035008 (2021)
6. Prigogine, I.: Introduction to Thermodynamics of Irreversible Processes. Interscience New York, New York (1955); de Groot, R., Mazur, P.: Non-equilibrium Thermodynamics. North-Holland Physics Publishing Amsterdam, Amsterdam (1961)
7. Breuer, H.-P., Petruccione, F.: The Theory of Open Quantum Systems. Oxford University Press, Oxford (2002)
8. Lindblad, G.: On the generators of quantum dynamical semigroups. Commun. Math. Phys. **48**, 119–130 (1976); Gorini, A.K., Sudarshan, E.C.G.: Completely positive dynamical semigroups of N-level systems. J. Math. Phys. **17**, 821–825 (1976)
9. Breuer, H.-P.: Quantum jumps and entropy production. Phys. Rev. A **68**, 032105 (2003)
10. Spohn, H.: Entropy production for quantum dynamical semigroups. J. Math. Phys. **19**, 1227 (1977)
11. Gardiner, C.: Handbook of Stochastic Methods. Springer, Berlin (2009)
12. Santos, J.P., Céleri, L.C., Landi, G.T., Paternostro, M.: The role of quantum coherence in non-equilibrium entropy production. npj Quant. Inf. **23**, 5 (2019)
13. Baumgratz, T., Cramer, M., Plenio, M.B.: Quantifying coherence. Phys. Rev. Lett. **113**, 140401 (2014)
14. Brunelli, M., Fusco, L., Landig, R., Wieczorek, W., Hoelscher-Obermaier, J., Landi, G.T., Semião, F.L., Ferraro, A., Kiesel, N., Donner, T., De Chiara, G., Paternostro, M.: Experimental determination of irreversible entropy production in out-of-equilibrium mesoscopic quantum systems. Phys. Rev. Lett. **121**, 160604 (2018)
15. Santos, J.P., Landi, G.T., Paternostro, M.: Wigner entropy production rate. Phys. Rev. Lett. **118**, 220601 (2017)
16. Santos, J.P., Céleri, L.C., Landi, G.T., Paternostro, M.: Spin-phase-space-entropy production. Phys. Rev. A **97**, 052123 (2018)
17. Wehrl, A.: General properties of entropy. Rev. Mod. Phys. **50**, 221–260 (1978); Wehrl: on the relation between classical and quantum-mechanical entropy. Rep. Math. Phys. **16**, 353–358 (1979)
18. Radcliffe, J.M.: Some properties of coherent spin states. J. Phys. A **4**, 313–323 (1971); Takahashi, F.S.: Spin coherent state representation in non-equilibrium statistical mechanics. J. Phys. Soc. Jpn. **38**, 656–668 (1975)
19. Zicari, G., Çakmak, B., Müstecaplıoğlu, Ö.E., Paternostro, M.: On the role of initial coherence in the spin phase- space entropy production rate. New J. Phys. **25**, 013030 (2023)

How Small Can Maxwell's Demon Be?

Lessons from Autonomous Electronic Feedback Models

Gernot Schaller

Abstract External piecewise-constant feedback control can modify energetic and entropic balances, allowing in extreme scenarios for Maxwell demon operational modes. Without specifying the actual implementation of external feedback loops, one can only partially quantify the additional contributions to entropy production. This is different in autonomously operating systems with internal feedback. Traditional (bipartite) autonomous systems can be divided into controller and a controlled subsystem, but also non-bipartite systems can accomplish the same task. We consider examples of autonomous three-terminal models that transfer heat mainly from a cold to a hot reservoir by dumping a small fraction of it to an ultra-cold (demon) reservoir, such that their coarse-grained dynamics resembles an external feedback loop. We find that the minimal three-level implementation is most efficient in utilizing heat dissipation to change the entropy balance of the effective controlled system.

1 Introduction

Maxwell's demon is a hypothetical being that in the original gedanken experiment [1] monitors the speed and direction of thermal particles and uses that information to spatially separate the particles into faster and slower fractions by opening and closing a shutter (which ideally requires no energy expenditure). This paradox has fascinated generations of researchers: Entropy is (locally) reduced and the resulting gradient could even be used to extract useful work, such that ideally the demon uses only information to lead to this apparent violation of the second law of thermodynamics. Formally speaking, it implements an external feedback control loop: A measurement reveals information about the particle (is it e.g. fast or slow, left or right-moving), the signal is then processed and the corresponding control operation is performed (shall the demon open or close the shutter).

G. Schaller (✉)
Helmholtz-Zentrum Dresden-Rossendorf, Bautzner Landstraße 400, 01328 Dresden, Germany
e-mail: g.schaller@hzdr.de

X. Bouju and C. Joachim (eds.), *Crossroad of Maxwell Demon*, Advances in Atom and Single Molecule Machines, https://doi.org/10.1007/978-3-031-57904-2_3

Nowadays, Maxwell's paradox is resolved by including the entropic balance of the demon in the discussion [2–4]. This already enters in effective descriptions of feedback control loops [5, 6] where the information gained from measurements can be experimentally quantified [7, 8]. Alternatively, one can represent the demon by a separate physical system [9, 10] that is in contact with the controlled one and study the energy and information flows within the bipartite joint system [11, 12]. It has been argued that the processing of information in this traditional sense is not a necessary ingredient [13, 14], which has fostered attempts to classify Maxwell demons more rigorously [15].

In any case, the actual nature of the feedback loop implementation becomes relevant, and in this book chapter contribution, we will compare how external feedback loops affect the energetic and entropic balance of a discrete (quantum) system with the situation in small autonomous systems that accomplish the same task.

2 Background: Energy/Particle-Resolving Rate Equations

The dynamics of many open systems with discrete quantum states can be described by rate equations—first order differential equations for probabilities P_a to occupy state a with energy E_a and particle number N_a—of the form

$$\dot{P}_a = \sum_\nu \sum_b R^\nu_{ab} P_b - \sum_\nu \sum_b R^\nu_{ba} P_a \,, \tag{1}$$

where $R^\nu_{ab} \geq 0$ is the transition rate from state b to a different state a due to reservoir ν (for stationary transport scenarios we require multiple reservoirs). Naturally, given a proper initial state, (1) obeys $P_a \in [0, 1]$ and $\sum_a P_a = 1$ at all times.[1]

Above, we have assumed that the transition rates are additively composed from different reservoirs. Such rate equations may arise as a special case of Lindblad master equations [16] and have their (limited) regime of validity, typically based on weak-coupling, Markovian and sometimes secular approximations. One should therefore keep in mind that with this formalism, results for the ultra-short regime, the strong-coupling regime, and the regime of near degenerate energies should be treated with caution, as the exact dynamics in these regimes may be much more complicated (e.g. non-Markovian or having non-additive rates). Furthermore, in the natural setting where one obtains these rate equations, one also assumes the reservoirs to be maintained at a thermal state, described by inverse temperature $\beta_\nu = (k_B T_\nu)^{-1}$ and possibly a chemical potential μ_ν. This eventually reflects in the fact that the transition rates for the same reservoir obey local detailed balance relations

[1] This is fulfilled when $R^\nu_{ab} \geq 0$ for $a \neq b$ and $R^\nu_{aa} = 0$ (these diagonal values would cancel out anyways). If we rewrite (1) in matrix form $\dot{\boldsymbol{P}} = \mathcal{R}\boldsymbol{P}$, the off-diagonal elements of the rate matrix will be given by the rates $\mathcal{R}_{a\neq b} = \sum_\nu R^\nu_{ab} \geq 0$, whereas the diagonal entries are such that the column sum of the rate matrix vanishes $\mathcal{R}_{aa} = -\sum_\nu \sum_{b\neq a} R^\nu_{ba}$, sufficient for probability conservation.

$$\frac{R^\nu_{ab}}{R^\nu_{ba}} = e^{-\beta_\nu[E_a - E_b - \mu_\nu(N_a - N_b)]}\,, \tag{2}$$

such that for $\mu_\nu = 0$ the energy-increasing transitions are suppressed compared to energy-decreasing transitions, and raising the chemical potentials tends to increase the system particle number. If the system is only coupled to a single reservoir or—equivalently—if all reservoirs are at the same inverse temperature β and chemical potential μ (such that we can drop the bath index ν), this can be used to show that the thermal state $\bar{P}_a = e^{-\beta(E_a - \mu N_a)}/Z$ is a stationary one.[2]

2.1 System Energy and Particle Balance

The time derivative of the system energy $\langle E\rangle = \sum_a E_a P_a$ can in absence of direct driving of the system be written as

$$\frac{d}{dt}\langle E\rangle = \sum_a E_a \dot{P}_a = \sum_\nu \sum_{ab} \left[E_a R^\nu_{ab} P_b - E_a R^\nu_{ba} P_a\right] = \sum_\nu \sum_{ab} (E_a - E_b) R^\nu_{ab} P_b\,, \tag{3}$$

where (by the additivity of the transition rates) the time-dependent energy current entering the system from reservoir ν becomes

$$I^\nu_E \equiv \sum_{ab} (E_a - E_b) R^\nu_{ab} P_b\,. \tag{4}$$

Neglecting the energy content in the system-reservoir interaction (weak-coupling limit), we can identify the reservoir changes with the negative energy current entering the system $dE_\nu/dt \approx -I^\nu_E$, and then the above just implies the first law of thermodynamics

$$\frac{d}{dt}\langle E\rangle + \sum_\nu \frac{dE_\nu}{dt} = 0\,, \tag{5}$$

indicating that the total energy of system and reservoirs together remains constant.[3] An analogous thought allows to construct the particle (matter) currents from the system particle number $\langle N\rangle = \sum_a N_a P_a$ via $\frac{d}{dt}\langle N\rangle \equiv \sum_\nu I^\nu_M$ with the time-dependent matter current entering the system from reservoir ν given by

---

[2] Consider $\sum_b R_{ab}\bar{P}_b - \sum_b R_{ba}\bar{P}_a = \sum_b R_{ab}/Z\left[e^{-\beta(E_b - \mu N_b)} - e^{+\beta[E_a - E_b - \mu(N_a - N_b)]}\, e^{-\beta(E_a - \mu N_a)}\right] = 0$. For reservoirs at different equilibrium states, the stationary state is a non-equilibrium one.

[3] Adding and subtracting the matter currents (6) we can also write the first law (5) as $\frac{d}{dt}\langle E\rangle = \sum_\nu \mu_\nu I^\nu_M + \sum_\nu (I^\nu_E - \mu_\nu I^\nu_M)$ in terms of a chemical work rate and heat currents, respectively.

$$I_M^\nu \equiv \sum_\nu \sum_{ab} (N_a - N_b) R_{ab}^\nu P_b \,. \tag{6}$$

To obtain the stationary currents $\bar{I}_{E/M}^\nu$ (we denote stationary values by an overbar throughout), one simply has to insert the stationary occupation probabilities for the system $P_b \to \bar{P}_b$. For a single thermal reservoir obeying (2), it follows that all currents at steady state must vanish, but this is no longer true for different reservoirs held at different temperatures and/or chemical potentials.

2.2 System Entropy Balance

The time derivative of the system's Shannon entropy $\langle S \rangle = -\sum_a P_a \ln P_a$ becomes[4]

$$\begin{aligned}
\frac{d}{dt}\langle S \rangle &= -\sum_a \dot{P}_a \ln P_a = -\sum_\nu \sum_{ab} R_{ab}^\nu P_b \ln P_a \frac{R_{ba}^\nu}{P_b R_{ab}^\nu} \frac{P_b R_{ab}^\nu}{R_{ba}^\nu} + \sum_\nu \sum_{ab} R_{ba}^\nu P_a \ln P_a \\
&= +\sum_\nu \sum_{ab} R_{ab}^\nu P_b \ln \frac{R_{ab}^\nu P_b}{R_{ba}^\nu P_a} + \sum_\nu \sum_{ab} R_{ab}^\nu P_b \ln \frac{R_{ba}^\nu}{R_{ab}^\nu} \frac{1}{P_b} + \sum_\nu \sum_{ab} R_{ab}^\nu P_b \ln P_b \\
&= +\sum_\nu \sum_{ab} R_{ab}^\nu P_b \ln \frac{R_{ab}^\nu P_b}{R_{ba}^\nu P_a} + \sum_\nu \sum_{ab} R_{ab}^\nu P_b \ln \frac{R_{ba}^\nu}{R_{ab}^\nu} \\
&= +\sum_\nu \sum_{ab} R_{ab}^\nu P_b \ln \frac{R_{ab}^\nu P_b}{R_{ba}^\nu P_a} + \sum_\nu \beta_\nu (I_E^\nu - \mu_\nu I_M^\nu) \,.
\end{aligned} \tag{7}$$

The first term on the r.h.s. of the last line is positive (which e.g. follows from the logarithmic sum inequality) and is known as (irreversible) entropy production rate [17, 18]

$$\dot{\sigma}_\mathrm{i} \equiv +\sum_\nu \sum_{ab} R_{ab}^\nu P_b \ln \frac{R_{ab}^\nu P_b}{R_{ba}^\nu P_a} \geq 0 \,. \tag{8}$$

The positivity of the entropy production rate quantifies the second law of thermodynamics

$$\dot{\sigma}_\mathrm{i} = \frac{d}{dt}\langle S \rangle - \sum_\nu \beta_\nu (I_E^\nu - \mu_\nu I_M^\nu) = \frac{d}{dt}\langle S \rangle + \sum_\nu \frac{dS_\nu}{dt} \geq 0 \,, \tag{9}$$

---

[4] In the topmost line of (7) we used $\sum_a \dot{P}_a = 0$, inserted the rate equation (1) and expanded the fraction in the first term of the r.h.s. In the second line we used properties of the logarithm and exchanged $a \leftrightarrow b$ in the last term. In the third line, we have combined the logarithms to insert the local detailed balance property (2) and identified the currents (4) and (6) in the last line.

where we have identified the second term on the r.h.s. with the entropy production rate of a reservoir without volume change maintained in thermal equilibrium.[5]

2.3 Stochastic Propagation

To solve (1), one may exponentiate the rate matrix $\mathcal{R}$, which is feasible when the dimension of $\mathcal{R}$ is small. Such a solution would only provide average quantities, whereas a system subject to continuous measurements of its energy and/or particle number will fluctuate, such that the rate equation solution $P_i(t)$ has to be reconstructed from many experimental trajectories. Such time-resolved measurements can be more directly compared to stochastic trajectory solutions.

These are generated as follows: Supposing that at time t, the system is in state a, one determines the random waiting time τ_a until the next quantum jump via

$$\tau_a = \frac{-\ln(1-r)}{\sum_\nu \sum_{b\neq a} R_{ba}^\nu}\,, \tag{10}$$

where r is a uniformly distributed random number with $r \in [0, 1]$. The waiting times determined this way are Poissonian-distributed according to the waiting-time distribution $\omega_a(\tau) = \sum_\nu \sum_{b\neq a} R_{ba}^\nu e^{-\sum_\nu \sum_{b\neq a} R_{ba}^\nu \tau}$. The system remains in the state a during $[t, t + \tau_a]$, and afterwards a quantum jump to a different state $b \neq a$ triggered by reservoir ν is performed with probability

$$P_{a\to b}^\nu = \frac{R_{ba}^\nu}{\sum_\mu \sum_{c\neq a} R_{ca}^\mu} \quad : \quad \sum_{b\neq a}\sum_\nu P_{a\to b}^\nu = 1\,, \tag{11}$$

such that after the waiting time, a jump is performed with certainty. The energetic and particle changes are attributed to reservoir ν, which allows to track the exchanged heat with the methods of stochastic thermodynamics [18, 19]. Then, one repeats the process with b being the new initial state.[6]

Additional continuous measurements of the system energy and particle number will have no additional effect on the dynamics.[7]

---

[5] To see that, solve $dE_\nu = T_\nu dS_\nu + \mu_\nu dN_\nu$ for the reservoir entropy change dS_ν/dt and use that for weak couplings we have $dE_\nu/dt \approx -I_E^\nu$ and $dN_\nu/dt = -I_M^\nu$.

[6] Numerically, the algorithm has the advantage that no rate matrix needs to be stored, no timestep width needs to be adjusted, just random numbers are required. To see that upon averaging trajectories one reproduces (1), consider that the waiting time distribution according to (10) just reproduces the probability to remain in state a via $P_{a\to a}(\tau) = 1 - \int_0^\tau \omega_a(t')dt' = e^{-\sum_\nu \sum_{b\neq a} R_{ba}^\nu \tau}$, and that the jump probability (11) is just the probability for jump $a \to b$ compliant with (1), conditioned on a jump actually happening.

[7] This is only true if the measurement basis coincides with the pointer basis of the rate equation, e.g., if we measure the total system energy E and total system particle number N. If we measure e.g.

2.4 Feedback Control

The trajectory formulation can be directly generalized to immediate and piecewise-constant external interventions conditioned on a perfect measurement of the systems state or its transitions: We assume that the control interventions only change the transition rates (e.g. via changing the system energies or the coupling strength to the reservoirs ν), but not the eigenstates of the quantum system themselves (which would require a treatment beyond rate equations [20]). The quality of this assumption depends of course on the applied experimental feedback protocol. Furthermore, we assume that the feedback operations are performed with negligible delay after the measurement (see e.g. Ref. [21] for delay effects) and that the control operations themselves leave the system particle number constant. Whereas ideal Maxwell-demon control leaves the system energy constant, more realistic control operations may change it [22], and this feedback work has to be taken into account.[8]

By convention, we label the energies that one would observe at a fixed moment in time (i.e., between the jumps) by E_a. These may differ from the energies in absence of feedback, as e.g. upon detection of the final state a, the control operation has established energy E_a. Accordingly, the energy difference $E_a - E_b$ for a system's jump between two states a and b may now be decomposed into heat, stochastically exchanged with the reservoirs, and feedback work, which is tied to the stochastic heat contributions, and this decomposition can be done in different ways.

For example, if we assume that the feedback is conditioned on detecting the system in a particular final state a with energy E'_a before the control operation (see Fig. 1 left panel), the feedback work cannot depend on how (e.g. by which reservoir) the system got there. This is different from the case where the feedback is conditioned on the detection of a quantum jump from $b \to a$ due to reservoir ν with energy exchange $E_a^\nu - E_b^\nu$: As the energy exchange may depend on the reservoir ν, so may the feedback work contribution, which may be further split into contributions prior and subsequent to the jump (Fig. 1 right panel). Another striking difference is that for state-conditioned feedback the heat transferred to and from the reservoir may differ for forward and backward transitions, whereas for jump-conditioned feedback it is the same (up to a sign).

To simulate merely the dynamics of the system, it is sufficient to replace the rates by their feedback-controlled versions $R_{ab}^\nu \to F_{ab}^\nu$ in (1), (10), and (11). To track the dynamics of heat and entropy, it is necessary to generalize the thermodynamic laws (5) and (9) to instantaneous and piecewise-constant feedback:

For state-conditioned feedback, we can in analogy to (4) write for the energy currents and the feedback work current

local particle numbers, the competition of different bases may induce additional quantum features like the quantum Zeno effect [20].

[8] As the feedback actions are conditioned on measurement results, the work rate for open-loop drivings $\dot{W}_{\rm drv} = \sum_a \dot{E}_a P_a$ does not apply here.

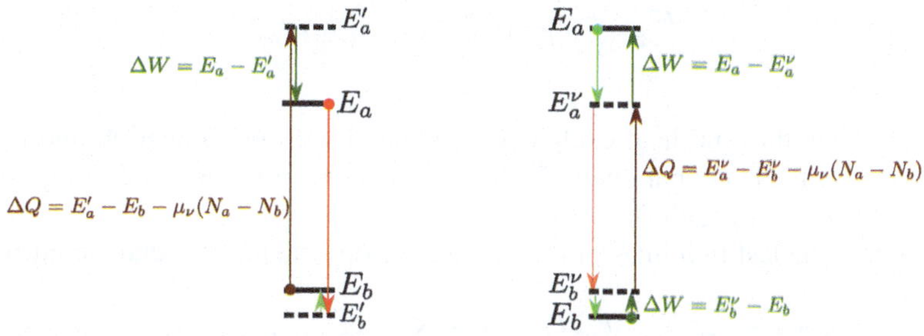

Fig. 1 Sketch of the instantaneous feedback scheme for the transition between states $b \to a$ (dark colors) and the reverse transition (light colors). Mere observation of the system energy at a fixed time would yield the levels E_a (solid black), but as the heat transfer (red) may not correspond to the exact energy differences, the auxiliary energies (dashed) are introduced to account for the feedback work (green). In the Maxwell-demon limit, we have $E_a^\nu \approx E_a$ and $E'_a \approx E_a$. **Left:** For state-conditioned feedback, the feedback work cannot depend on the reservoir ν. **Right:** For jump-conditioned feedback, the feedback work is split into an energetic contribution preceding the jump and one succeeding it

$$I_E^\nu = \sum_{ab:a\neq b} (E'_a - E_b) F^\nu_{ab} P_b \,, \qquad I_E^{\text{fb}} = \sum_\nu \sum_{ab:a\neq b} (E_a - E'_a) F^\nu_{ab} P_b \,, \tag{12}$$

where we have used that every control operation is tied to a change in states, such that the corresponding energy difference is multiplied with the corresponding transition probability. The total energy change (first law of thermodynamics) is then obtained by summing

$$\frac{d}{dt}\langle E\rangle = \sum_\nu I_E^\nu + I_E^{\text{fb}} = \sum_\nu \sum_{ab} (E_a - E_b) F^\nu_{ab} P_b \,, \tag{13}$$

which reflects the fact that the system is switching between energies E_a, see Fig. 1. For jump-conditioned feedback, we proceed similarly and write (4) as

$$I_E^\nu = \sum_{ab:a\neq b} (E^\nu_a - E^\nu_b) F^\nu_{ab} P_b \,, \qquad I_E^{\text{fb}} = \sum_\nu \sum_{ab:a\neq b} (E^\nu_b - E_b + E_a - E^\nu_a) F^\nu_{ab} P_b \,, \tag{14}$$

and their sum yields the same result as before. So far, these partitions into heat and feedback work contributions are somewhat arbitrary, they can be fixed using some microscopic intuition as we do in Sects. 3.1 and 3.2.

Even when the feedback does not change the system energies (e.g., when only the transition rates are modified), the local detailed balance relation (2) will no longer hold. To account for this and the proper corresponding energetic exchanges with the reservoirs, we parametrize the deviation from local detailed balance as [5]

$$\frac{F^\nu_{ab}}{F^\nu_{ba}} = e^{-\beta_\nu[\Delta E^\nu_{ab} - \mu_\nu(N_a - N_b)]} e^{+\delta^\nu_{ab}}, \tag{15}$$

where ΔE^ν_{ab} is the true heat exchange associated with the transition and δ^ν_{ab} is a dimensionless feedback parameter.[9] The feedback-free situation is thus reproduced for $\Delta E^\nu_{ab} = E_a - E_b$ and $\delta^\nu_{ab} \to 0$.

With this, the last two lines of the system entropy balance (7) change into

$$\begin{aligned}\frac{d}{dt}\langle S\rangle &= +\sum_\nu \sum_{ab} F^\nu_{ab} P_b \ln \frac{F^\nu_{ab} P_b}{F^\nu_{ba} P_a} + \sum_\nu \sum_{ab:a\neq b} F^\nu_{ab} P_b \ln \left[e^{\beta_\nu[\Delta E^\nu_{ab} - \mu_\nu(N_a - N_b)} e^{-\delta^\nu_{ab}}\right] \\ &= +\sum_\nu \sum_{ab} F^\nu_{ab} P_b \ln \frac{F^\nu_{ab} P_b}{F^\nu_{ba} P_a} + \sum_\nu \beta_\nu (I^\nu_E - \mu_\nu I^\nu_M) - \sum_\nu \sum_{ab:a\neq b} F^\nu_{ab} P_b \delta^\nu_{ab}\,.\end{aligned} \tag{16}$$

As before, the first term is positive by construction and is identified with the entropy production rate of the (model) universe, the second term is still interpreted as the entropy production in the reservoirs (which are untouched by the feedback discussed here), but the last term quantifies how the feedback changes the entropy balance of the system and can thus be identified with an information current

$$I_I \equiv +\sum_\nu \sum_{ab:a\neq b} F^\nu_{ab} P_b \delta^\nu_{ab}, \tag{17}$$

such that the second law (9) generalizes to

$$\sigma_\mathrm{i} = \frac{d}{dt}\langle S\rangle - \sum_\nu \beta_\nu (I^\nu_E - \mu_\nu I^\nu_M) + I_I \geq 0\,. \tag{18}$$

Even at steady state (where $\frac{d}{dt}\langle S\rangle \to 0$), it is now possible to have negative entropy produced in the reservoirs ($-\sum_\nu \beta_\nu (I^\nu_E - \mu_\nu I^\nu_M) \leq 0$) as long as the information current counter-balances it such that the above modified second law (18) is still respected. Of course, this treatment neglects a (possibly large) part of the entropy produced in the unspecified feedback loop, it only takes into account its effect on the system. Maxwell demon feedbacks, in which we are mostly interested here, generate a situation where the feedback energy current is negligible $\left|I^\mathrm{fb}_E\right| \ll \left|I^\nu_E\right|$, while at the same time, the information current is substantial $\mathcal{O}\{I_I\} = \mathcal{O}\{\beta_\nu (I^\nu_E - \mu_\nu I^\nu_M)\}$. Both feedback schemes are capable of generating such scenarios.

[9] Note that one has the relation $\delta^\nu_{ab} + \delta^\nu_{ba} = \beta_\nu (\Delta E^\nu_{ab} + \Delta E^\nu_{ba})$.

2.5 Maxwell-Demon Feedback for a Two-Level System

When we consider a two-level system P_0, P_1 with energies $E_0 < E_1$ coupled to two (hot and cold) thermal reservoirs $\nu \in \{h, c\}$, the populations of its energy eigenstates follow in absence of feedback the rate equation

$$\frac{d}{dt}\begin{pmatrix} P_0 \\ P_1 \end{pmatrix} = \begin{pmatrix} -R_{10}^c - R_{10}^h & +R_{01}^c + R_{01}^h \\ +R_{10}^c + R_{10}^h & -R_{01}^c - R_{01}^h \end{pmatrix} \begin{pmatrix} P_0 \\ P_1 \end{pmatrix}, \tag{19}$$

where by construction, we assume the implicit condition $T_c < T_h$ or $\beta_c > \beta_h$. The rates satisfy local detailed balance (2).[10] To check thermodynamic consistency, consider the stationary energy current entering the system via the cold junction $\bar{I}_E^c = (E_1 - E_0)\frac{R_{10}^c R_{01}^h - R_{01}^c R_{10}^h}{R_{10}^c + R_{01}^c + R_{10}^h + R_{01}^h} = (E_1 - E_0)\frac{R_{10}^c R_{10}^h (e^{\beta_h(E_1 - E_0)} - e^{\beta_c(E_1 - E_0)})}{R_{10}^c + R_{01}^c + R_{10}^h + R_{01}^h}$. This is always negative, which means that heat flows from hot to cold (Clausius formulation of the second law). This is also confirmed by looking at the entropy production rate (9), which at steady state (where $\bar{I}_M^h = -\bar{I}_M^c$ and $\bar{I}_E^h = -\bar{I}_E^c$ and the system contribution is negligible) simplifies to $\dot{\bar{\sigma}}_{\rm i} = (\beta_h - \beta_c)\bar{I}_E^c \geq 0$.

In presence of feedback, we now consider an increase or a decrease (compared to the feedback-free case) of the bare tunnelling rate to reservoir ν parametrized by feedback parameters δ_0^ν and δ_1^ν, such that the feedback-controlled rates relate to the feedback-free rates as

$$\begin{aligned} F_{10}^c &= R_{10}^c e^{+\delta_0^c}, & F_{01}^c &= R_{01}^c e^{+\delta_1^c}, \\ F_{10}^h &= R_{10}^h e^{+\delta_0^h}, & F_{01}^h &= R_{01}^h e^{+\delta_1^h}, \end{aligned} \tag{20}$$

and we assume the energies to remain unperturbed $\Delta E_{ab}^\nu = E_a - E_b$, such that $\delta_{01}^\nu = \delta_1^\nu - \delta_0^\nu = -\delta_{10}^\nu$. This can significantly alter the transport characteristics.[11] Accordingly, we just replace $R_{ij}^\nu \to F_{ij}^\nu$ in the rate matrix (19). Then, the stationary energy current entering the system from the cold junction becomes (for Maxwell-demon type feedbacks we have by energy conservation $\bar{I}_E^h = -\bar{I}_E^c$)

$$\begin{aligned} \bar{I}_E^c &= (E_1 - E_0)\frac{F_{10}^c F_{01}^h - F_{01}^c F_{10}^h}{F_{10}^c + F_{01}^c + F_{10}^h + F_{01}^h} \\ &= (E_1 - E_0)\frac{R_{10}^c R_{10}^h (e^{\beta_h(E_1 - E_0)} e^{\delta_0^c + \delta_1^h} - e^{\beta_c(E_1 - E_0)} e^{\delta_1^c + \delta_0^h})}{F_{10}^c + F_{01}^c + F_{10}^h + F_{01}^h}, \end{aligned} \tag{21}$$

---

[10] Often occurring rates of this form are Bose distributions $R_{10}^\nu = \Gamma_\nu [e^{\beta_\nu(E_1 - E_0)} - 1]^{-1}$ and $R_{01}^\nu = \Gamma_\nu \{1 + [e^{\beta_\nu(E_1 - E_0)} - 1]^{-1}\}$ or Fermi functions $R_{10}^\nu = \Gamma_\nu [e^{\beta_\nu(E_1 - E_0)} + 1]^{-1}$ and $R_{01}^\nu = \Gamma_\nu \{1 - [e^{\beta_\nu(E_1 - E_0)} + 1]^{-1}\}$, where the bare rates Γ_ν would cancel in (2).

[11] Consider the extreme limit $\delta_0^c \to +\infty$, $\delta_1^c \to -\infty$, $\delta_0^h \to -\infty$, and $\delta_1^h \to +\infty$, where the two-level system couples only to the cold reservoir when it is in its ground state and to the hot reservoir while it is in its excited state. In consequence, the energy can only flow from cold to hot reservoir.

which is positive if and only if $\delta_0^c + \delta_1^h - \delta_1^c - \delta_0^h = \delta_{01}^h - \delta_{01}^c > (\beta_c - \beta_h)(E_1 - E_0)$. Thus, with feedback we can direct the stationary energy current from the cold reservoir to the hot one by using sufficiently strong feedback parameters δ_a^ν. The first law reads as before (Maxwell demon feedback by assumption does not inject energy), but the information current modifies the second law, such that the entropy production rate under feedback (18) becomes manifestly positive

$$\dot{\bar{\sigma}}_{\rm i} = (\beta_h - \beta_c)\bar{I}_E^c + \bar{I}_I \geq 0 \quad : \quad \bar{I}_I = \frac{\delta_0^c - \delta_1^c - \delta_0^h + \delta_1^h}{E_1 - E_0}\bar{I}_E^c = \frac{\delta_{01}^h - \delta_{01}^c}{E_1 - E_0}\bar{I}_E^c .$$

3 Autonomous Systems

The implementation of external feedback control loops is experimentally challenging [23] and has shortcomings in its theoretical discussion (effective thermodynamic discussion, highly idealized assumptions on measurement and feedback speed, ...). On-chip implementations of feedback loops in higher-dimensional rate equations do not suffer from these problems but allow for convenient cycle analyses [24]. Indeed, ideal Maxwell demons can – at least within the limits of perturbative descriptions – be very closely approximated [25]. Aiming at smallest systems, we focus here on setups that can direct the flow of energy from a cold reservoir towards a hot one by sacrificing a small fraction of the energy to a ultra-cold third terminal, see Fig. 2.

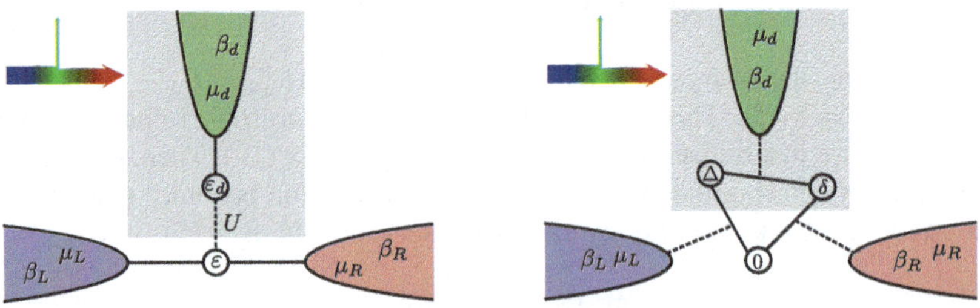

Fig. 2 Autonomous demonic refrigerators considered in the subsections below. Solid lines denote tunnel junctions whereas dashed lines indicate merely energetic exchanges. With three (blue left, red right, and green demon) terminals at inverse temperatures $\beta_d > \beta_L > \beta_R$ (chemical potentials can be chosen to vanish throughout $\mu_\nu = 0$), it is possible to cool the left (blue) reservoir. In specific regimes, the devices act like autonomous Maxwell demons in the sense that in ignorance of the demon reservoir (shaded region), heat flows from cold (blue) to hot (red) with negligible energy loss (green) across the demon junction (arrows on top left). **Left**: The implementation with two capacitively interacting quantum dots (circles) with variable particle number leads to a bipartite system (bottom) - controller (shaded) setup with four levels. **Right**: An analog non-bipartite three-level system can be implemented by the singly-charged sector of three quantum dots with tunnelling events between them triggered by the reservoirs

3.1 Autonomous 4-Level System: Interacting Dots

We first review a bipartite model shown in Fig. 2 left panel, that has been discussed theoretically with focus on battery charging [10] and also has been implemented experimentally [26]. It consists of two interacting two-level subsystems (quantum dots), such that our total system has four levels with the corresponding occupation probabilities P_{00}, P_{10}, P_{01}, P_{11} of finding the first system in state 0, 1, 0, 1 while the second is in states 0, 0, 1, 1, respectively. The first system is tunnel-coupled to two (left and right) reservoirs with additive transition rates, whereas the second is only coupled to one reservoir. Due to Coulomb interaction, the transition rates of one subsystem may depend on the state of the other. The rate matrix $\mathcal{R}$ becomes

$$\begin{pmatrix} -\gamma_\uparrow^d - \gamma_\uparrow^L - \gamma_\uparrow^R & \gamma_\downarrow^L + \gamma_\downarrow^R & \gamma_\downarrow^d & 0 \\ \gamma_\uparrow^L + \gamma_\uparrow^R & -\gamma_\uparrow^{dU} - \gamma_\downarrow^L - \gamma_\downarrow^R & 0 & \gamma_\downarrow^{dU} \\ \gamma_\uparrow^d & 0 & -\gamma_\downarrow^d - \gamma_\uparrow^{LU} - \gamma_\uparrow^{RU} & \gamma_\downarrow^{LU} + \gamma_\downarrow^{RU} \\ 0 & \gamma_\uparrow^{dU} & \gamma_\uparrow^{LU} + \gamma_\uparrow^{RU} & -\gamma_\downarrow^{dU} - \gamma_\downarrow^{LU} - \gamma_\downarrow^{RU} \end{pmatrix} .$$

The rates can be written (beyond the wideband approximation) as products of bare tunneling rates $\Gamma_\nu^{(U)}$ and Fermi functions $f_\nu^{(U)}$

$$\begin{aligned} \gamma_\uparrow^\nu &= \Gamma_\nu f_\nu , & \gamma_\downarrow^\nu &= \Gamma_\nu (1 - f_\nu) , \\ \gamma_\uparrow^{\nu U} &= \Gamma_\nu^U f_\nu^U , & \gamma_\downarrow^{\nu U} &= \Gamma_\nu^U (1 - f_\nu^U) , \end{aligned} \tag{22}$$

where the left and right Fermi functions are given by $f_\nu = [e^{\beta_\nu(\varepsilon - \mu_\nu)} + 1]^{-1}$ and $f_\nu^U = [e^{\beta_\nu(\varepsilon + U - \mu_\nu)} + 1]^{-1}$ for the system reservoirs $\nu \in \{L, R\}$ and analogously $f_d = [e^{\beta_d(\varepsilon_d - \mu_d)} + 1]^{-1}$ and $f_d^U = [e^{\beta_d(\varepsilon_d + U - \mu_d)} + 1]^{-1}$ for the demon reservoir.

3.1.1 Function: Cooling the Left Reservoir

To implement cooling on the left reservoir while $\beta_L > \beta_R$ and $\mu_L = \mu_R$, one needs a cold demon reservoir and energies and chemical potentials tuned such that $f_d \to 1$ and $f_d^U \to 0$, and furthermore some asymmetry in the left and right tunnelling rates, such that $\Gamma_L^U, \Gamma_R \gg \Gamma_L, \Gamma_R^U$. A typical trajectory in this operational regime is depicted in Fig. 3. Since the device operates stochastically, one has unsuccessful loops like $|01\rangle \to |00\rangle \to |01\rangle$ (which has zero net energy exchange), but in the chosen regime, successful loops like $|01\rangle \to |11\rangle \to |10\rangle \to |00\rangle \to |01\rangle$ become likely. This loop effectively transfers energy $\varepsilon + U$ out of the left reservoir to the demon reservoir with fraction U and the right reservoir with fraction ε. Thus, the regime of a proper Maxwell demon is reached when we have $U \ll \varepsilon$ while the directionality of all transitions is maintained. The required conditions on parameters $\Gamma_d, \Gamma_d^U \gg \Gamma_{L/R}, \Gamma_{L/R}^U$ and $\Gamma_L^U, \Gamma_R \gg \Gamma_L, \Gamma_R^U$ and $\beta_d U \gg 1$ can be simultaneously fulfilled [10].

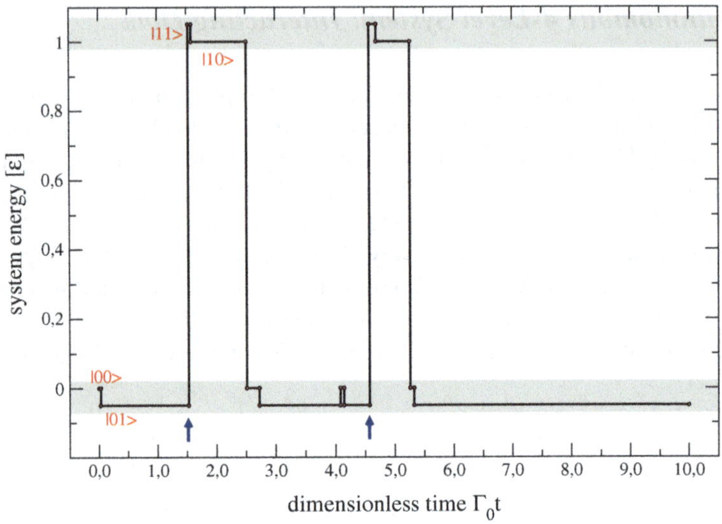

Fig. 3 Typical trajectory (see Sect. 2.3) for the four-level system in the operational regime of cooling functionality for the left reservoir. The system is initialized in state $|00\rangle$, and right afterwards the controller dot is loaded (changing its energy to ε_d chosen negative here) leading to state $|01\rangle$. Then, occasionally the dot is loaded from the left (the right tunneling rate at this energy is suppressed) reservoir (transferring energy $\varepsilon + U$ into the system, cooling functionality). However, for a successful cycle (blue arrows), the controller dot unloads quickly to reach state $|10\rangle$ (dissipating heat $\varepsilon_d + U$ into the demon reservoir), the system can then unload an electron with energy ε to the right (the left tunnelling rate at this energy is suppressed), effectively transferring energy ε to the right reservoir. The cycle is then closed by the controller dot loading again leading to state $|01\rangle$. Shaded regions mark the coarse-grained states of the system dot (23). Other parameters: $\Gamma_L^U = \Gamma_R = \Gamma_0$, $\Gamma_d = \Gamma_d^U = 10\Gamma_0, \Gamma_L = \Gamma_U^R = 0, \varepsilon_d = -.05\varepsilon, U = 0.1\varepsilon, \mu_\nu = 0, \beta_d\varepsilon = 100, \beta_R\varepsilon = 1, \beta_L\varepsilon = 2$

3.1.2 Relation to the Feedback-Controlled Model

We can introduce the marginal probabilities for the state of the controlled system by summing over the states of the demon dot

$$P_0 = P_{00} + P_{01}\,, \qquad P_1 = P_{10} + P_{11}\,, \tag{23}$$

and when constructing the differential equations for them we rewrite $P_{ij} = \frac{P_{ij}}{P_i} P_i \approx (\lim_{\Gamma_d, \Gamma_d^U \to \infty} \frac{P_{ij}}{P_i}) P_i$. The in reality time-dependent prefactor can in the limit of a very fast demon reservoir be replaced by a constant

$$\lim_{\Gamma_d\to\infty}\frac{P_{00}}{P_0}=\frac{\gamma_\downarrow^d}{\gamma_\downarrow^d+\gamma_\uparrow^d}=1-f_d\,,\qquad \lim_{\Gamma_d\to\infty}\frac{P_{01}}{P_0}=\frac{\gamma_\uparrow^d}{\gamma_\downarrow^d+\gamma_\uparrow^d}=f_d\,,$$
$$\lim_{\Gamma_d^U\to\infty}\frac{P_{10}}{P_1}=\frac{\gamma_\downarrow^{dU}}{\gamma_\downarrow^{dU}+\gamma_\uparrow^{dU}}=1-f_d^U\,,\qquad \lim_{\Gamma_d^U\to\infty}\frac{P_{11}}{P_1}=\frac{\gamma_\uparrow^{dU}}{\gamma_\downarrow^{dU}+\gamma_\uparrow^{dU}}=f_d^U\,, \tag{24}$$

where $0\le f_d\le 1$ and $0\le f_d^U\le 1$, to obtain a coarse-grained rate equation [18, 27, 28] for the probabilities P_0, P_1 of the system dot being empty or filled like shown by the shaded regions of Fig. 3, respectively. In the limit where $f_d\to 1$ and $f_d^U\to 0$, the coarse-grained rate matrix simplifies[12] further into

$$\mathcal{R}_{\rm cg}=\begin{pmatrix}-(\gamma_\uparrow^{LU}+\gamma_\uparrow^{RU}) & (\gamma_\downarrow^L+\gamma_\downarrow^R)\\ +(\gamma_\uparrow^{LU}+\gamma_\uparrow^{RU}) & -(\gamma_\downarrow^L+\gamma_\downarrow^R)\end{pmatrix}. \tag{25}$$

In this, we see that the effective transition energy for the relaxation rate is different from the transition energy for the excitation, such that the autonomous four-level system effectively implements a state-conditioned feedback discussed in Sect. 2.4. The relevant ratios of effective transition rates for our entropic balance can be written as

$$\frac{F_{01}^\nu}{F_{10}^\nu}\to\frac{\gamma_\downarrow^\nu}{\gamma_\uparrow^{\nu U}}=\frac{\gamma_\downarrow^\nu}{\gamma_\uparrow^\nu}\frac{\gamma_\uparrow^\nu}{\gamma_\uparrow^{\nu U}}=e^{\beta_\nu(\varepsilon-\mu_\nu)}\frac{\gamma_\uparrow^\nu}{\gamma_\uparrow^{\nu U}}\,,$$
$$\frac{F_{10}^\nu}{F_{01}^\nu}\to\frac{\gamma_\uparrow^{\nu U}}{\gamma_\downarrow^\nu}=\frac{\gamma_\uparrow^{\nu U}}{\gamma_\downarrow^{\nu U}}\frac{\gamma_\downarrow^{\nu U}}{\gamma_\downarrow^\nu}=e^{-\beta_\nu(\varepsilon+U-\mu_\nu)}\frac{\gamma_\downarrow^{\nu U}}{\gamma_\downarrow^\nu}\,, \tag{26}$$

where we have used that the heat transfer for an initially filled system dot (and empty demon dot) is ε, whereas it is $\varepsilon+U$ for an initially empty system (and filled demon dot). From this, we can identify with (15) the energy differences $\Delta E_{01}^\nu=-\varepsilon$ and $\Delta E_{10}^\nu=\varepsilon+U$, which can be satisfied[13] with energies $E_0=0=E_0'$, and $E_1=\varepsilon$ and $E_1'=\varepsilon+U$. The feedback parameters become

$$\delta_{01}^\nu=\ln\frac{\gamma_\uparrow^\nu}{\gamma_\uparrow^{\nu U}}\,,\qquad \delta_{10}^\nu=\ln\frac{\gamma_\downarrow^{\nu U}}{\gamma_\downarrow^\nu}\,, \tag{27}$$

such that $\delta_{01}^\nu+\delta_{10}^\nu=\beta_\nu U$.

[12] This happens at low demon reservoir temperatures $\beta_d U\gg 1$ with e.g. $\varepsilon_d\approx\mu_d-U/2$. It is also the limit where the measurement becomes perfect and system and demon dot are always anti-correlated.

[13] These shifts can be understood as unloading of the dot transfers energy ε to the left or right lead, whereas loading requires energy $\varepsilon+U$ (the difference is transferred to the demon reservoir as effective feedback work).

We can thus write for the coarse-grained energy currents

$$I_{E,\rm cg}^{\nu} = (\varepsilon + U)F_{10}^{\nu}P_0 - \varepsilon F_{01}^{\nu}P_1\,, \qquad I_E^{\rm fb} = -U(F_{01}^{L} + F_{01}^{R})P_1\,, \tag{28}$$

These currents match what is obtained from the microscopic four-level model in the limit of a fast and precise demon.

Accordingly, the information current becomes

$$I_I = \sum_{\nu}\left[F_{01}^{\nu}P_1 \ln\frac{\gamma_{\uparrow}^{\nu}}{\gamma_{\uparrow}^{\nu U}} + F_{10}^{\nu}P_0 \ln\frac{\gamma_{\downarrow}^{\nu U}}{\gamma_{\downarrow}^{\nu}}\right]. \tag{29}$$

As the coarse-grained entropy production $\dot{\bar{\sigma}}_{\rm i,cg} = -\beta_L \bar{I}_{E,\rm cg}^{L} - \beta_R \bar{I}_{E,\rm cg}^{R} + \bar{I}_I \le \dot{\bar{\sigma}}_{\rm i}$ under-estimates the full entropy production [27] $\dot{\bar{\sigma}}_{\rm i} = -\beta_L \bar{I}_E^L - \beta_R \bar{I}_E^R - \beta_d \bar{I}_E^d$, and the coarse-grained energy currents converge to the ones from the microscopic model $\bar{I}_{E,\rm cg}^{\nu} \to \bar{I}_E^{\nu}$, it follows that the information current is smaller than the entropy production in the controller reservoir, such that the information gained in the measurement is only partially used in the feedback loop. Furthermore, it can be concluded that when we evaluate the thermodynamic uncertainty relation [29] $\dot{\bar{\sigma}}_{\rm i}\frac{\bar{S}}{\bar{I}^2} \ge 2$, (where $\bar{I}$ and $\bar{S}$ are energy or matter current and its associated noise over an arbitrary junction), with the coarse-grained entropy production instead, we can find regimes where it is violated.

3.2 Autonomous Three-Level System

We now consider a three-level system as in Fig. 2 right panel, with its three transitions selectively driven by three independent reservoirs. Such a system could be the single-charged sector of a triple quantum dot (see e.g. Ref. [30] for an experimental implementation). We assume that the left (cold) reservoir drives the large transition between ground and second excited state, the right (hot) reservoir the smaller transition between ground and first excited state, and the demon reservoir (coldest) the remaining one between first and second excited state.[14] We consider here the limit of a low-temperature demon reservoir with negligible energetic contribution that reverses the usual heat flow. In the basis P_0, P_δ, P_Δ, the rate matrix for our model reads

[14] This is different from the commonly used quantum absorbtion refrigerator (QAR) [31, 32]: First, in a QAR the hot reservoir drives the large excitation above the ground state. Second, a QAR would require the transition between the two excited states driven by a work reservoir (e.g. a reservoir of substantially higher temperature than the hot and cold ones). Third, the operational regime of a QAR is characterized by a non-negligible energy current from the work reservoir $\bar{I}_E^{w} = \bar{I}_E^{L}/\kappa \ge \bar{I}_E^{L}/\kappa_{\rm Ca}$, with the cofficient of performance $\kappa \le \kappa_{\rm Ca}$ limited by its Carnot value $\kappa_{\rm Ca} = \beta_R/(\beta_L - \beta_R)$. Unlike a Maxwell demon, a QAR is thus energy-dominated, it could be implemented by reverting the temperature hierarchy to $\beta_d < \beta_L < \beta_R$.

$$\mathcal{R} = \begin{pmatrix} -\gamma_\uparrow^L - \gamma_\uparrow^R & \gamma_\downarrow^R & \gamma_\downarrow^L \\ \gamma_\uparrow^R & -\gamma_\downarrow^R - \gamma_\uparrow^d & \gamma_\downarrow^d \\ \gamma_\uparrow^L & \gamma_\uparrow^d & -\gamma_\downarrow^L - \gamma_\downarrow^d \end{pmatrix}, \tag{30}$$

where $\gamma_\uparrow^\nu$ denotes the excitation and $\gamma_\downarrow^\nu$ the de-excitation rate due to reservoir ν.

3.2.1 Function: Cooling the Left Reservoir

In the desired operational mode, the second excited state is occasionally reached from the ground state by the absorbtion of energy Δ from the cold left reservoir (implementing cooling). The ultra-cold demon reservoir then takes a fraction $\Delta - \delta$ of that energy to transfer the system from the second to its first excited state. Finally, the hot reservoir absorbs heat δ by allowing the transfer from the first excited state back to the ground state, closing the cycle. The opposite process is blocked by the ultra-cold demon reservoir in the intended regime, see Fig. 4 for a typical trajectory. Thus, for a successful cycle $|0\rangle \to |\Delta\rangle \to |\delta\rangle \to |0\rangle$, energy Δ is taken from the left reservoir, while energies $\Delta - \delta$ and δ are dissipated into the demon and right reservoirs, respectively. The Maxwell demon limit is thus approached when $\delta \to \Delta$, while the directionality of transitions is maintained. For our model, this requires $\Gamma^d \gg \Gamma^L, \Gamma^R$ and $\beta_d(\Delta - \delta) \gg 1$.

The stationary currents entering the system from the reservoirs read

$$\bar{I}_E^L = \frac{\Delta(\gamma_\uparrow^L \gamma_\downarrow^d \gamma_\downarrow^R - \gamma_\downarrow^L \gamma_\uparrow^d \gamma_\uparrow^R)}{[\gamma^L \gamma^d - \gamma_\downarrow^L \gamma_\downarrow^d] + [\gamma^L \gamma^R - \gamma_\uparrow^L \gamma_\uparrow^R] + [\gamma^R \gamma^d - \gamma_\downarrow^R \gamma_\uparrow^d]}, \tag{31}$$

where $\gamma^\nu \equiv \gamma_\uparrow^\nu + \gamma_\downarrow^\nu$. The other currents are tightly coupled $\bar{I}_E^R = -\frac{\delta}{\Delta}\bar{I}_E^L$ and $\bar{I}_E^d = -\frac{\Delta-\delta}{\Delta}\bar{I}_E^L$, such that the sum of all stationary currents vanishes (first law) and by using $\gamma_\uparrow^d \to 0$ we can implement cooling of the cold left reservoir ($\bar{I}_E^L > 0$).

At steady-state, the entropy production rate (9) thus reads

$$\dot{\sigma}_{\rm i} = -\beta_L \bar{I}_E^L - \beta_R \bar{I}_E^R - \beta_d \bar{I}_E^d = \left(\beta_R \frac{\delta}{\Delta} + \beta_d \frac{\Delta - \delta}{\Delta} - \beta_L\right) \bar{I}_E^L \geq 0. \tag{32}$$

Thus, cooling occurrs when the prefactor in round brackets is positive, i.e., when $\beta_d - \beta_L \geq (\beta_d - \beta_R)\delta/\Delta$. Furthermore, by choosing $\Delta - \delta \to 0$ the energy dissipated into the demon reservoir can be made arbitrarily small while the entropy produced there remains large due to its low temperature $\beta_d(\Delta - \delta) \gg 1$. This already indicates a demon operational mode,[15] but we can make the relation to the external control loop more explicit by applying a similar coarse-graining procedure as before.

---

[15] An efficiency of the device could be defined by $\eta = -\bar{I}_E^R/\bar{I}_E^L \Theta(\bar{I}_E^L) = \delta/\Delta\Theta(\bar{I}_E^L) \leq 1 - (\beta_L - \beta_R)/(\beta_d - \beta_R)$ which is upper-bounded by unity as $\beta_d \to \infty$.

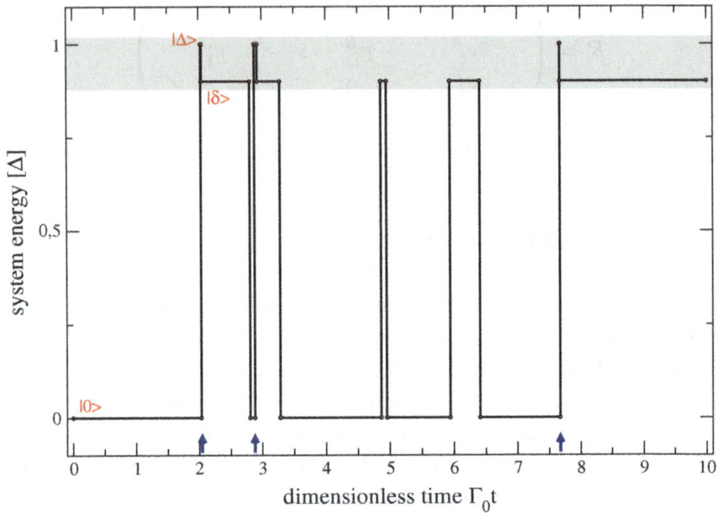

Fig. 4 Typical trajectory (see Sect. 2.3) for the three-level system in the operational regime of cooling functionality for the left reservoir. The system is initialized in state $|0\rangle$, and is occasionally excited to $|\delta\rangle$ by the right, or to $|\Delta\rangle$ by the left reservoir. In the first case, the excitation is mostly given back to the right reservoir as the demon reservoir has a much lower temperature that prohibits further excitation (e.g. middle of the plot). From state $|\Delta\rangle$ though (blue arrows), the demon reservoir quickly induces a relaxation to state $|\delta\rangle$, from where via the transition to $|0\rangle$ most of the heat is then dissipated into the right reservoir (thereby closing the cooling cycle). The shaded region marks the coarse-grained states (33). Other parameters: $\gamma_\uparrow^\nu = \Gamma_\nu n_\nu$ and $\gamma_\downarrow^\nu = \Gamma_\nu(1+n_\nu)$ with Bose distributions $n_L = [e^{\beta_L \Delta} - 1]^{-1}, n_R = [e^{\beta_R \delta} - 1]^{-1}, n_d = [e^{\beta_d(\Delta-\delta)} - 1]^{-1}$ and $\Gamma_L = \Gamma_R = \Gamma_0$, $\Gamma_d = 100\Gamma_0$, $\delta = 0.9\Delta$, $\beta_d = 100\beta_R$, $\beta_L = 2\beta_R$, $\beta_R\Delta = 1$

3.2.2 Relation to the Feedback-Controlled Model

In the limit where the demon reservoir rates are both significantly larger than the others while maintaining local detailed balance (2) (consider e.g. $\gamma_\downarrow^d = \Gamma_d(1+n_d)$, $\gamma_\uparrow^d = \Gamma_d n_d$ with Bose distribution $n_d = [e^{\beta_d(\Delta-\delta)} - 1]^{-1}$ in the limit $\Gamma_d \to \infty$), we can formally coarse-grain (a similar argument is used in Sect. V of Ref. [33]) the rate equation by introducing the probability to be in any of the excited states[16] as indicated by the shaded region in Fig. 4

$$P_1 = P_\delta + P_\Delta\,. \tag{33}$$

Then, the coarse-graining approximation can be implemented by using that the conditional probabilities of being in one of the excited states are given by

[16] Technically, we construct the equation for $\dot{P}_1$ from the rate equation and rewrite $P_\delta = (P_\delta/P_1)P_1 \to [\gamma_\downarrow^d/(\gamma_\downarrow^d + \gamma_\uparrow^d)]P_1$ and $P_\Delta = (P_\Delta/P_1)P_1 \to [\gamma_\uparrow^d/(\gamma_\downarrow^d + \gamma_\uparrow^d)]P_1$ in all equations.

$$\lim_{\gamma^d_{\uparrow/\downarrow}\to\infty} \frac{P_\delta}{P_1} = \frac{\gamma^d_\downarrow}{\gamma^d_\downarrow + \gamma^d_\uparrow} = \frac{1}{1+e^{-\beta_d(\Delta-\delta)}}\,, \qquad \lim_{\gamma^d_{\uparrow/\downarrow}\to\infty} \frac{P_\Delta}{P_1} = \frac{\gamma^d_\uparrow}{\gamma^d_\downarrow + \gamma^d_\uparrow} = \frac{1}{1+e^{+\beta_d(\Delta-\delta)}}\,,$$

such that the coarse-grained rate matrix for the states (P_0, P_1) becomes

$$\mathcal{R}_{\rm cg} = \begin{pmatrix} -\gamma_\uparrow^L - \gamma_\uparrow^R & +\gamma_\downarrow^L[1+e^{+\beta_d(\Delta-\delta)}]^{-1} + \gamma_\downarrow^R[1+e^{-\beta_d(\Delta-\delta)}]^{-1} \\ +\gamma_\uparrow^L + \gamma_\uparrow^R & -\gamma_\downarrow^L[1+e^{+\beta_d(\Delta-\delta)}]^{-1} - \gamma_\downarrow^R[1+e^{-\beta_d(\Delta-\delta)}]^{-1} \end{pmatrix}. \tag{34}$$

Since the ratios of the $\gamma_\downarrow^\nu$ and $\gamma_\uparrow^\nu$ already give the proper heat exchanges with those reservoirs, the three-level model thus effectively implements jump-conditioned feedback from Sect. 2.4. From (15) we can directly identify the feedback parameters as

$$\delta_{01}^L = \ln\frac{1}{1+e^{+\beta_d(\Delta-\delta)}} = -\delta_{10}^L\,, \qquad \delta_{01}^R = \ln\frac{1}{1+e^{-\beta_d(\Delta-\delta)}} = -\delta_{10}^R\,, \tag{35}$$

such that $\delta_{01}^\nu + \delta_{10}^\nu = 0$. In the zero-temperature demon limit, the cold reservoir is thus associated with divergent (infinitely strong) feedback.

The stationary energy current from the left reservoir becomes

$$\bar{I}_{E,\rm cg}^L = -\Delta F_{01}^L \bar{P}_{12} + \Delta F_{10}^L \bar{P}_0 = \Delta \frac{\gamma_\uparrow^L\gamma_\downarrow^R - e^{-\beta_d(\Delta-\delta)}\gamma_\downarrow^L\gamma_\uparrow^R}{\gamma_\uparrow^L + \gamma_\uparrow^R + \gamma_\downarrow^R + e^{-\beta_d(\Delta-\delta)}(\gamma_\downarrow^L + \gamma_\uparrow^L + \gamma_\uparrow^R)}\,, \tag{36}$$

such that cooling is possible under the same conditions as before. The energy current from the hot reservoir is fixed by the tight coupling condition $\bar{I}_{E,\rm cg}^R = -\frac{\delta}{\Delta}\bar{I}_{E,\rm cg}^L$ and also the energy current injected by the feedback becomes $\bar{I}_E^{\rm fb} = -\frac{\Delta-\delta}{\Delta}\bar{I}_{E,\rm cg}^L$ (as one may verify). These currents match what is obtained for the microscopic three-level model in the limit $\Gamma_d \to \infty$ for the left, right, and controller reservoir, respectively.

Accordingly, the entropy produced in the demon reservoir is given by $\dot{\sigma}_{\rm i}^d = \beta_d \frac{\Delta-\delta}{\Delta}\bar{I}_E^L$. This matches the effective information current resulting from the feedback, which becomes

$$\bar{I}_I = \sum_\nu \delta_{01}^\nu (F_{01}^\nu \bar{P}_1 - F_{10}^\nu \bar{P}_0) = \frac{\beta_d(\Delta-\delta)}{\Delta}\bar{I}_E^L\,. \tag{37}$$

The three-level implementation is thus more efficient than the four-level version not only by its reduced complexity, but also its more efficient use of information: The information current due to the demon precisely corresponds to the entropy produced in the feedback loop. The drawback is that there is no parameter to turn off the feedback smoothly (like U was used in the four-level model).

3.2.3 Electronic Implementation

One possibility to implement an electronic three-level system is in the Coulomb-blockade regime of two Coulomb-coupled quantum dots with effective single-particle onsite energies Δ and δ. Preferentially, the left (cold) reservoir should load and unload a dot mode with charging energy Δ, the right (hot) reservoir should load and unload a mode with charging energy δ. The selectivity of the coupling may be realized by using peaked reservoir spectral functions (which may be generated or enhanced further by connecting over additional fine-tuned quantum dots [34]) resonant with the corresponding transition energies Δ and δ, respectively. The remaining transition between the two singly-charged modes is made possible by the action of a third (demon) reservoir, which leaves the particle number in the system constant.[17]

If one aims at setups where one has only energy but no particle exchange, one would need three particle-conserving reservoirs. Then, an electronic implementation of a three-level system would be possible using the singly charged sector of three tunnel-coupled quantum dots $H_S^1 = \sum_{ij} \tau_{ij} d_i^\dagger d_j$ coupled to three particle-conserving reservoirs. An original interaction coupling to $S_\nu \propto d_i^\dagger d_i$ will then generate transition operators and energy renormalizations for all triple dot modes. Also here, the selectivity of the interaction could be implemented by tuning the reservoir parameters accordingly.

The rates for the demon reservoir in the double-dot case or all reservoirs in the singly-charged sector of the triple dot case can be microscopically calculated [35]. To see that, consider a fermionic reservoir $H_B = \sum_k \varepsilon_k c_k^\dagger c_k$ with a coupling of the form

$$H_I = S \otimes \left[\sum_{kq} t_{kq} c_k^\dagger c_q - \sum_k t_{kk} f(\varepsilon_k)\right] \equiv S \otimes B\,, \tag{38}$$

where S is a generic system coupling operator, and $t_{kq} = t_{qk}^*$ describes the reservoir-internal scattering[18] and the term with the Fermi function $f(\varepsilon_k) = [e^{\beta(\varepsilon_k-\mu)} + 1]^{-1}$ has been introduced to cancel the expectation value of the coupling operator at equilibrium.[19] The correlation function $C(\tau) = \mathrm{Tr_B}\left\{e^{+\mathrm{i}H_B\tau} B e^{-\mathrm{i}H_B\tau} B \bar{\rho}_B\right\}$ can then be written as a two-fold integral $C(\tau) = \frac{1}{2\pi}\int d\omega \int d\omega' T(\omega,\omega') f(\omega)[1-f(\omega')] e^{\mathrm{i}(\omega-\omega')\tau}$, where we have introduced the

---

[17] Any capacitive interaction with coupling operator $S = u_1 d_1^\dagger d_1 + u_2 d_2^\dagger d_2$ will trigger transitions between eigenmodes of the system $S = \alpha_\delta c_\delta^\dagger c_\delta + \alpha_\Delta c_\Delta^\dagger c_\Delta + [\tau c_\Delta^\dagger c_\delta + \text{h.c.}]$.

[18] A toy model for such a reservoir could be a chain with N modes (where the actual reservoir limit is obtained for $N \to \infty$) $H_B = \varepsilon \sum_{i=1}^N d_i^\dagger d_i + \tau \sum_{i=1}^{N-1} (d_i^\dagger d_{i+1} + d_{i+1}^\dagger d_i)$ with homogeneous on-site energies ε and tunnelling amplitudes τ. For this, one can use the unitary transformation $d_i = \sum_k \sqrt{2/(N+1)} \sin[\pi i k/(N+1)] c_k$ to diagonalize it exactly into $H_B = \sum_{k=1}^N \varepsilon_k c_k^\dagger c_k$ with $\varepsilon_k = \varepsilon + 2\tau \cos[\pi k/(N+1)]$. When we couple this reservoir at its first site to a system via $H_I = S \otimes [d_1^\dagger d_1 - \langle d_1^\dagger d_1 \rangle]$, we would obtain $t_{kq} = 2/(N+1) \cdot \sin[\pi k/(N+1)] \sin[\pi q/(N+1)]$.

[19] Without renormalization, such an interaction would perturb the equilibrium state of the reservoir.

function $T(\omega, \omega') = 2\pi \sum_{kq} \left|t_{kq}\right|^2 \delta(\omega - \varepsilon_k)\delta(\omega' - \varepsilon_q) = T(\omega', \omega)$ that describes the scattering processes inside the reservoir.[20] Independent of the chemical potential μ (which may be used to tune the overall magnitude), the Fourier transform $\gamma(\Omega) \equiv \int C(\tau)e^{+i\Omega\tau}d\tau = \int d\omega T(\omega, \omega + \Omega) f(\omega)[1 - f(\omega + \Omega)]$ obeys the KMS relation $\gamma(-\Omega) = \gamma(+\Omega)e^{-\beta\Omega}$, which implies that the associated rates $R_{ab} = \gamma(E_b - E_a)|\langle a| S |b\rangle|^2$ obey the detailed balance condition (2) with $\mu_\nu \to 0$ and $\beta_\nu \to \beta$. In order to have not too small rates, the actual chemical potential should be tuned near the desired transition energy. Furthermore, energetic filtering for the chain toy model is possible by choosing ε also resonant with the desired transition energy and small τ. Further simplifications are possible when the scattering function is approximately constant.[21]

Unfortunately, particularly in the interesting Maxwell demon limit where $\delta \to \Delta$ one has two near-degenerate excitation energies in the system, and energetic filtering alone is insufficient. In this case, one can additionally tune the system and reservoir couplings to maintain the selectivity of the reservoirs.[22]

4 Conclusions

To answer the main question: Maxwell's demon, incarnated by autonomous demonic refrigerators, can be be pretty small (our considerations suggested three levels). In addition, it does not require a bipartite system-controller setup and can also be very efficient in the sense that the entropy produced in the feedback loop is equivalent to the information current entering the effective controlled system.

The present contribution has focused on the mathematical framework of rate equations with the associated limitations. Generalizations to equations describing situations of near-degenerate energies [36], with system-reservoir correlations [37], and the strong-coupling regime [38] in autonomous systems or not perfectly-energetic measurements [20, 39] and delay effects [40] in external loops are interesting pathways of further research.

---

[20] For the chain model reservoir it would for $N \to \infty$ assume the form $T(\omega, \omega') = 2/(\pi\tau^2) \cdot \sqrt{1 - (\omega - \varepsilon)^2/(4\tau^2)}\sqrt{1 - (\omega' - \varepsilon)^2/(4\tau^2)}\Theta(4\tau^2 - (\omega - \varepsilon)^2)\Theta(4\tau^2 - (\omega' - \varepsilon)^2)$, which shows that by tuning ε and τ one may modify the spectral function of the reservoir.

[21] When $T(\omega, \omega + \Omega) \approx T(\mu, \mu) \equiv T_\mu$, we obtain $\gamma(\Omega) \to T_\mu \Omega[1 + n_B(\Omega)]$ with Bose distribution $n_B(\Omega) = [e^{\beta\Omega} - 1]^{-1}$, leading to a transition rate $R_{ab} = T_\mu |\langle a| S |b\rangle|^2 (E_b - E_a) [1 + n_B(E_b - E_a)]$.

[22] For a homogeneous triple quantum dot $H_S = \varepsilon_S[d_1^\dagger d_1 + d_2^\dagger d_2 + d_3^\dagger d_3] - \tau_S[d_1^\dagger d_2 + d_2^\dagger d_3 + d_3^\dagger d_1 + \text{h.c.}]$ one finds with $d_1 = (c_1 - \sqrt{2}c_3)/\sqrt{3}$, $d_{2,3} = (\sqrt{2}c_1 \mp \sqrt{3}c_2 + c_3)/\sqrt{6}$ a representation in terms of two degenerate eigenmodes $H_S = [\varepsilon_S - 2\tau_S]c_1^\dagger c_1 + [\varepsilon_S + \tau_S][c_2^\dagger c_2 + c_3^\dagger c_3]$. Then, (up to dephasing terms $\propto c_i^\dagger c_i$) the capacitive coupling $S_c \propto d_1^\dagger d_1$ drives transitions between states $1 \leftrightarrow 3$, $S_h \propto d_1^\dagger d_1 + 2d_3^\dagger d_3$ drives transitions between $1 \leftrightarrow 2$ (desired) and $2 \leftrightarrow 3$ (can be filtered), and finally $S_d \propto d_2^\dagger d_2$ drives all possible transitions (but $1 \leftrightarrow 2$ and $1 \leftrightarrow 3$ can be filtered). This can be further improved when non-homogeneous tripledot systems are considered.

Acknowledgements The author thanks C. Kohlfürst and P. Strasberg for helpful comments on the manuscript and for generous conference support by the CRC 1242 (Project ID No. 278162697).

References

1. Maxwell, J.C.: Theory of Heat. Greens, and Co., London, Longmans (1871)
2. Leff, H., Rex, A.: Maxwell's Demon: Entropy, Information, Computing. Princeton University Press, Princeton, New Jersey (1990)
3. Leff, H., Rex, A.: Maxwell's Demon 2: Entropy, Classical and Quantum Information. Computing. CRC Press, Bristol (2002)
4. Maruyama, K., Nori, F., Vedral, V.: Rev. Mod. Phys. **81**(1), 1 (2009). https://doi.org/10.1103/RevModPhys.81.1
5. Esposito, M., Schaller, G.: Europhys. Lett. **99**, 30003 (2012). https://doi.org/10.1209/0295-5075/99/30003
6. Bergli, J., Galperin, Y.M., Kopnin, N.B.: Phys. Rev. E **88**, 062139 (2013). http://link.aps.org/doi/10.1103/PhysRevE.88.062139
7. Vidrighin, M.D., Dahlsten, O., Barbieri, M., Kim, M.S., Vedral, V., Walmsley, I.A.: Phys. Rev. Lett. **116**, 050401 (2016). https://doi.org/10.1103/PhysRevLett.116.050401
8. Cottet, N., Jezouin, S., Bretheau, L., Campagne-Ibarcq, P., Ficheux, Q., Anders, J., Aufféves, A., Azouit, R., Rouchond, P., Huard, B.: Proc. Natl. Acad. Sci. U. S. A. **114**, 7561 (2017). https://doi.org/10.1073/pnas.1704827114
9. Mandal, D., Quan, H.T., Jarzynski, C.: Phys. Rev. Lett. **111**, 030602 (2013). https://doi.org/10.1103/PhysRevLett.111.030602
10. Strasberg, P., Schaller, G., Brandes, T., Esposito, M.: Phys. Rev. Lett. **110**, 040601 (2013). https://doi.org/10.1103/PhysRevLett.110.040601
11. Horowitz, J.M., Esposito, M.: Phys. Rev. X **4**, 031015 (2014). https://doi.org/10.1103/PhysRevX.4.031015
12. Ehrich, J., Sivak, D.A.: Front. Phys. **11**, 1108357 (2023). https://doi.org/10.3389/fphy.2023.1108357
13. Sánchez, R., Splettstoesser, J., Whitney, R.S.: Phys. Rev. Lett. **123**, 216801 (2019). https://doi.org/10.1103/PhysRevLett.123.216801
14. Ciliberto, S.: Phys. Rev. E **102**, 050103 (2020). https://doi.org/10.1103/PhysRevE.102.050103
15. Freitas, N., Esposito, M.: Phys. Rev. E **103**, 032118 (2021). https://doi.org/10.1103/PhysRevE.103.032118
16. Breuer, H.P., Petruccione, F.: The Theory of Open Quantum Systems. Oxford University Press, Oxford (2002)
17. Esposito, M., Harbola, U., Mukamel, S.: Phys. Rev. E **76**(3), 031132 (2007). https://doi.org/10.1103/PhysRevE.76.031132
18. Strasberg, P.: Quantum Stochastic Thermodynamics (Oxford University Press, Oxford, 2022)
19. Seifert, U.: Rep. Prog. Phys. **75**, 126001 (2012). https://doi.org/10.1088/0034-4885/75/12/126001
20. Schaller, G., Cerrillo, J., Engelhardt, G., Strasberg, P.: Phys. Rev. B **97**, 195104 (2018). https://doi.org/10.1103/PhysRevB.97.195104
21. Emary, C.: Philos. Trans. R. Soc. A **371**, 1471 (2013). https://doi.org/10.1098/rsta.2012.0468
22. Ehrich, J., Still, S., Sivak, D.A.: Phys. Rev. Res. **5**, 023080 (2023). https://doi.org/10.1103/PhysRevResearch.5.023080
23. Chida, K., Desai, S., Nishiguchi, K., Fujiwara, A.: Nat. Commun. **8**, 15310 (2017). https://doi.org/10.1038/ncomms15301
24. Mayrhofer, R.D., Elouard, C., Splettstoesser, J., Jordan, A.N.: Phys. Rev. B **103**, 075404 (2021). https://doi.org/10.1103/PhysRevB.103.075404

25. Sánchez, R., Samuelsson, P., Potts, P.P.: Phys. Rev. Res. **1**, 033066 (2019). https://doi.org/10.1103/PhysRevResearch.1.033066
26. Koski, J.V., Kutvonen, A., Khaymovich, I.M., Ala-Nissila, T., Pekola, J.P.: Phys. Rev. Lett. **115**, 260602 (2015). https://doi.org/10.1103/PhysRevLett.115.260602
27. Esposito, M.: Phys. Rev. E **85**, 041125 (2012). https://doi.org/10.1103/PhysRevE.85.041125
28. Ehrich, J.: J. Stat. Mech.: Theory Exp. **2021**(8), 083214 (2021). https://doi.org/10.1088/1742-5468/ac150e
29. Barato, A.C., Seifert, U.: Phys. Rev. Lett. **114**, 158101 (2015). https://doi.org/10.1103/PhysRevLett.114.158101
30. Seo, M., Choi, H.K., Lee, S.Y., Kim, N., Chung, Y., Sim, H.S., Umansky, V., Mahalu, D.: Phys. Rev. Lett. **110**, 046803 (2013). https://doi.org/10.1103/PhysRevLett.110.046803
31. Levy, A., Kosloff, R.: Phys. Rev. Lett. **108**, 070604 (2012). https://doi.org/10.1103/PhysRevLett.108.070604
32. Linden, N., Popescu, S., Skrzypczyk, P.: Phys. Rev. Lett. **105**(13), 130401 (2010). https://doi.org/10.1103/PhysRevLett.105.130401
33. Strasberg, P., Schaller, G., Brandes, T., Jarzynski, C.: Phys. Rev. E **90**, 062107 (2014). https://doi.org/10.1103/PhysRevE.90.062107
34. Ehrlich, T., Schaller, G.: Phys. Rev. B **104**, 045424 (2021). https://doi.org/10.1103/PhysRevB.104.045424
35. Schaller, G.: Open Quantum Systems Far from Equilibrium. Lecture Notes in Physics, vol. 881. Springer, Cham (2014)
36. Trushechkin, A.: Phys. Rev. A **103**, 062226 (2021). https://doi.org/10.1103/PhysRevA.103.062226
37. Beyer, K., Luoma, K., Strunz, W.T.: Phys. Rev. Lett. **123**, 250606 (2019). https://doi.org/10.1103/PhysRevLett.123.250606
38. Strasberg, P., Schaller, G., Schmidt, T.L., Esposito, M.: Phys. Rev. B **97**, 205405 (2018). https://doi.org/10.1103/PhysRevB.97.205405
39. Seah, S., Nimmrichter, S., Scarani, V.: Phys. Rev. Lett. **124**, 100603 (2020). https://doi.org/10.1103/PhysRevLett.124.100603
40. Debiossac, M., Grass, D., Alonso, J.J., Lutz, E., Kiesel, N.: Nat. Commun. **11**, 1360 (2020). https://doi.org/10.1038/s41467-020-15148-5

Many-Body Thermal States on a Quantum Computer: A Variational Approach

Mirko Consiglio and Tony J. G. Apollaro

Abstract Many-body quantum states at thermal equilibrium are ubiquitous in nature. Investigating their dynamical properties is a formidable task due to the complexity of the Hilbert space they live in. Quantum computers may have the potential to effectively simulate quantum systems, provided that the many-body state under scrutiny can be faithfully prepared via an efficient algorithm. With this aim, we present a hybrid quantum–classical variational quantum algorithm for the preparation of the Gibbs state of the quantum XY model. Our algorithm is based on the Grover and Rudolph parametrized quantum circuit for the preparation of the Boltzmann weights of the Gibbs state, and on a parity-preserving ansatz for the allocation of the eigenenergy basis to their respective Boltzmann weight. We explicitly show, with a paradigmatic few-body case instance, how the symmetries of a many-body system can be exploited to significantly reduce the exponentially increasing number of variational parameters needed in the Grover and Rudolph algorithm. Finally, we show that the density matrix, of the Gibbs state of the XY model, obtained by statevector simulations for different parameters, exhibits a fidelity close to unity with the exact Gibbs state; this highlights the potential use of our protocol on current quantum computers.

1 Introduction

Quantum many-body systems present formidable challenges due to the inherent complexity of the Hilbert space they encompass [1]. Even the task of finding the ground-state of such systems is generally categorized as an NP-hard problem [2]. A recent and promising approach for tackling computational problems that are computationally demanding on classical computers is to harness the power of quantum

M. Consiglio · T. J. G. Apollaro (✉)
Department of Physics, University of Malta, Msida 2080, Malta
e-mail: tony.apollaro@um.edu.mt

M. Consiglio
e-mail: mirko.consiglio@um.edu.mt

X. Bouju and C. Joachim (eds.), *Crossroad of Maxwell Demon*, Advances in Atom and Single Molecule Machines, https://doi.org/10.1007/978-3-031-57904-2_4

computers [3]. However, it is important to note that the current state of quantum computation is constrained to the capabilities achievable on NISQ (Noisy Intermediate-Scale Quantum) computers [4]. These machines are limited in that they can only apply quantum circuits of limited depth to a few noisy qubits before issues such as decoherence and gate errors compromise the accuracy of the obtained results.

In the past decade, a hybrid quantum–classical paradigm has been proposed as a means to harness the best of both classical and NISQ quantum computation [5]. These quantum–classical algorithms delegate the classically intractable part of an algorithm to the quantum computer, while performing efficiently subroutines on a classical computer, such as objective function optimization. A paradigmatic example of such procedures are variational quantum algorithms (VQAs) [6], where a parametrized quantum circuit (PQC), along with the classical optimizer, is tasked to prepare the quantum state encoding the solution to a given problem [7]. VQAs have been applied to a variety of quantum many-body problems, ranging from determining the electronic structure of small molecules [8], investigating the onset of exotic phases of matter [9], to detecting entanglement [10].

At the same time, one of the main applications of quantum computing is expected to be the simulation of quantum many-body systems. In particular, simulating a quantum system at thermal equilibrium is key for several disciplines, including quantum chemistry, biology and combinatorial optimisation. Evidently, preparing a quantum many-body system at thermal equilibrium is QMA-hard [11]. Different quantum algorithms have been proposed for the preparation of thermal states based on fluctuation theorems [12], imaginary time evolution [13], utilizing a generalized quantum assisted simulator [14] and via adiabatic quantum computation [15].

In this chapter we illustrate a thermal state preparation algorithm for a quantum many-body system based on the VQA first proposed in Ref. [16], where the PQC acts on an ancillary and a system register on which, respectively, the Boltzmann distribution and the Gibbs state are prepared. The advantage of this approach is that the von Neumann entropy, which is not an observable, can be determined without truncation. We apply our algorithm to the XY spin-$\frac{1}{2}$ Heisenberg Hamiltonian utilizing the Grover and Rudolph algorithm [17] for preparing the Boltzmann distribution, highlighting how the symmetries of the model can be utilized to reduce the number of variational parameters in the circuit, followed by using a parity-preserving ansatz for diagonalizing the Hamiltonian. The chapter is organized as follows: in Sect. 2 we introduce the XY Hamiltonian of which we aim to prepare the thermal state; in Sect. 3 we present the Grover and Rudolph algorithm used to prepare probability distribution functions on a quantum computer; in Sect. 4 we implement the symmetries of the model in order to reduce the number of variational parameters used for the thermal state preparation; in Sect. 5 we present our VQA and its performance in preparing the XY thermal state; finally, in Sect. 6 we draw future prospects of our approach.

2 The Model

In this Section we present the quantum many-body system of which we will prepare the thermal equilibrium state via a VQA. For the sake of completeness, we include a thorough derivation of its exact solution, which, due to the quadratic nature of the Hamiltonian in the fermionic representation, can be achieved both for finite and infinite system sizes.

Let us consider a 1D spin-$\frac{1}{2}$ system with periodic boundary conditions exhibiting nearest-neighbor interactions of Heisenberg type in the XY plane, subject to a uniform magnetic field in the transverse z direction, see Fig. 1 for a finite-size instance. The XY spin-$\frac{1}{2}$ Hamiltonian [18] is given by

$$\hat{H} = -J \sum_{n=1}^{N} \left((1+\gamma)\, \hat{S}_n^x \hat{S}_{n+1}^x + (1-\gamma)\, \hat{S}_n^y \hat{S}_{n+1}^y + \frac{h}{2} \hat{S}_n^z \right), \tag{1}$$

where $\hat{S}_i^\alpha$ ($\alpha = x, y, z$) is the spin-$\frac{1}{2}$ operator sitting on site i of a 1D lattice composed of N even particles. The anisotropy parameter $\gamma \in [0, 1]$ defines the XX model ($\gamma = 0$), the Ising ($\gamma = 1$) and the XY model ($0 < \gamma < 1$), with the last two belonging to the same universality class. In the following, the diagonalization is carried out for arbitrary γ and h.

One can rewrite Eq. (1) in terms of Pauli matrices $\hat{S}_i^\alpha = \frac{\hat{\sigma}_i^\alpha}{2}$

$$\hat{H} = -\frac{J}{2} \sum_{n=1}^{N} \left(\frac{1+\gamma}{2} \hat{\sigma}_n^x \hat{\sigma}_{n+1}^x + \frac{1-\gamma}{2} \hat{\sigma}_n^y \hat{\sigma}_{n+1}^y + \frac{h}{2} \hat{\sigma}_n^z \right), \tag{2}$$

and turn to the spin ladder operators $\hat{\sigma}^\pm = \frac{\hat{\sigma}^x \pm i \hat{\sigma}^y}{2}$, resulting in

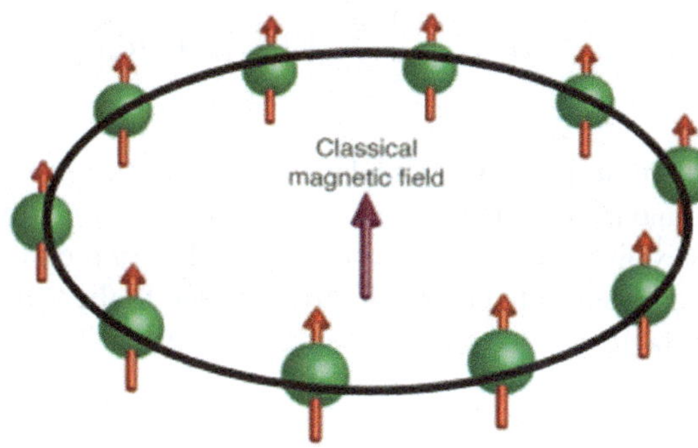

Fig. 1 Sketch of a 1D spin-$\frac{1}{2}$ XY Heisenberg Hamiltonian with periodic boundary conditions in the presence of a magnetic field as modeled by Eq. (2)

$$\hat{H} = -\frac{J}{2}\sum_{n=1}^{N}\left(\hat{\sigma}_n^+\hat{\sigma}_{n+1}^- + \hat{\sigma}_n^-\hat{\sigma}_{n+1}^+ + \gamma\left(\hat{\sigma}_n^+\hat{\sigma}_{n+1}^+ + \hat{\sigma}_n^-\hat{\sigma}_{n+1}^-\right)\right) - Jh\sum_{n=1}^{N}\hat{\sigma}_n^+\hat{\sigma}_n^- + \frac{NJh}{2}\hat{\mathbb{1}}$$
$$= -\frac{1}{2}\sum_{n=1}^{N}\left(\hat{\sigma}_n^+\hat{\sigma}_{n+1}^- + \hat{\sigma}_n^-\hat{\sigma}_{n+1}^+ + \gamma\left(\hat{\sigma}_n^+\hat{\sigma}_{n+1}^+ + \hat{\sigma}_n^-\hat{\sigma}_{n+1}^-\right)\right) - h\sum_{n=1}^{N}\hat{\sigma}_n^+\hat{\sigma}_n^-, \tag{3}$$

where in the last steps $[\hat{\sigma}^+, \hat{\sigma}^-] = \hat{\sigma}^z$ and $\{\hat{\sigma}^+, \hat{\sigma}^-\} = \hat{\mathbb{1}}$ have been used. We have also set $J = 1$ and rescaled the Hamiltonian by the constant term $\frac{NJh}{2}$. Applying the Jordan-Wigner transformation:

$$\hat{c}_n = e^{-i\pi\sum_{m=1}^{n-1}\hat{\sigma}_m^+\hat{\sigma}_m^-}\hat{\sigma}_n^-, \; \hat{c}_n^\dagger = \hat{\sigma}_n^+ e^{i\pi\sum_{m=1}^{n-1}\hat{\sigma}_m^+\hat{\sigma}_m^-}, \tag{4}$$

the Hamiltonian in the fermionic representation reads

$$\hat{H} = -\frac{1}{2}\sum_{n=1}^{N-1}\left(\hat{c}_n^\dagger\hat{c}_{n+1} + \gamma\hat{c}_n^\dagger\hat{c}_{n+1}^\dagger + \text{h.c.}\right) - h\sum_{n=1}^{N}\hat{c}_n^\dagger\hat{c}_n - \hat{P}\left(\hat{c}_N^\dagger\hat{c}_1 + \hat{c}_N^\dagger\hat{c}_1^\dagger + \text{h.c}\right), \tag{5}$$

where h.c. denotes the hermitian conjugate and the last term on the RHS accounts for the periodic boundary conditions, by defining the Parity operator $\hat{P} = \prod_{i=1}^{N}\left(-\hat{\sigma}_n^z\right)$. The Parity operator $\hat{P}$ has eigenvalues ± 1, denoting, respectively, the positive and the negative parity subspace. Hence, introducing the projectors on the positive and negative parity subspaces $\hat{P}^{\pm} = \frac{1}{2}\left(1 \pm \hat{P}\right)$, Eq. (5) can be cast in a direct sum of disjoint subspaces

$$\hat{H} = \hat{H}^+ + \hat{H}^-, \text{ where } \hat{H}^{\pm} = \hat{P}^{\pm}\hat{H}\hat{P}^{\pm}, \tag{6}$$

and

$$\hat{H}^{\pm} = -\frac{1}{2}\sum_{n=1}^{N}\left(\hat{c}_n^\dagger\hat{c}_{n+1} + \gamma\hat{c}_n^\dagger\hat{c}_{n+1}^\dagger + \text{h.c.}\right) - h\sum_{n=1}^{N}\hat{c}_n^\dagger\hat{c}_n, \tag{7}$$

where antiperiodic boundary conditions, $\hat{c}_{N+1} = -\hat{c}_1$, hold for $\hat{H}^+$, and periodic boundary conditions, $\hat{c}_{N+1} = \hat{c}_1$, hold for $\hat{H}^-$. While parity effects may become relevant for finite-size systems [19, 20], in the thermodynamic limit $N \to \infty$, they can be neglected being of order $\mathcal{O}\left(\frac{1}{N}\right)$. In the following we will diagonalize Eq. (6) in the respective parity subspaces for finite size systems.

2.1 Positive Parity Subspace Diagonalization

Because of the translational invariance of the system, we can perform a Fourier transformation of the fermionic operators,

$$\hat{c}_n^\dagger = \frac{e^{\frac{i\pi}{4}}}{\sqrt{N}} \sum_{k\in\mathcal{K}^+} e^{ikn} \hat{c}_k^\dagger, \ \hat{c}_n = \frac{e^{\frac{-i\pi}{4}}}{\sqrt{N}} \sum_{k\in\mathcal{K}^+} e^{-ikn} \hat{c}_k, \tag{8}$$

where the momenta $k \in \mathcal{K}^+$ are chosen in order to satisfy the antiperiodic boundary conditions for $\hat{H}^+$, and $\mathcal{K}^+ = -\frac{N-1}{N}\pi + m\frac{2\pi}{N}$, with $m = 0, 1, \ldots, N-1$ integers. The factor $e^{\frac{i\pi}{4}}$ has been introduced so as to end up with a real Hamiltonian, reading

$$\hat{H}^+ = \sum_{k\in\mathcal{K}^+} (h - \cos k) \left(\hat{c}_k^\dagger \hat{c}_k + \hat{c}_{-k}^\dagger \hat{c}_{-k}\right) - \gamma \sin k \left(\hat{c}_k^\dagger \hat{c}_{-k}^\dagger + \hat{c}_{-k} \hat{c}_k\right). \tag{9}$$

The spinless, quadratic Hamiltonian in Eq. (9) is finally diagonalized via a Bogoliubov transformation

$$\hat{c}_k = \cos\frac{\theta_k}{2}\hat{\gamma}_k - \sin\frac{\theta_k}{2}\hat{\gamma}_{-k}^\dagger, \text{ where } \tan\theta_k = \frac{\gamma \sin k}{h - \cos k}, \tag{10}$$

yielding

$$\hat{H}^+ = \sum_{k\in\mathcal{K}^+} \epsilon_k \left(\hat{\gamma}_k^\dagger \hat{\gamma}_k - \frac{1}{2}\right), \tag{11}$$

where the single-particle energy spectrum is given by

$$\epsilon_k = \sqrt{(h - \cos k)^2 + \gamma^2 \sin^2 k}. \tag{12}$$

The ground-state of $\hat{H}^+$ is given by the vacuum of the $\hat{\gamma}_k$ fermions and has energy

$$E_0^+ = -\frac{1}{2} \sum_{k\in\mathcal{K}^+} \epsilon_k, \tag{13}$$

and the full spectrum of $\hat{H}^+$ is derived by adding an even number of fermions to the ground-state

$$E_k^+ = E_0^+ + \sum_{k\in\mathcal{P}_{\text{even}}(\mathcal{K}^+)} \epsilon_k, \tag{14}$$

where $\mathcal{P}_{\text{even}}(\mathcal{K}^+)$ is the subset of the power set of $\mathcal{K}^+$ with an even number of terms.

2.2 Negative Parity Subspace Diagonalization

The diagonalization of the model in the negative parity subspace follows closely that in the positive one and is sketched below. In order to fulfill the periodic boundary condition for the negative parity subspace Hamiltonian $\hat{H}^-$, the Fourier transformation is carried over $k \in \mathcal{K}^- = -\pi + (m+1)\frac{2\pi}{N}$ with $m = 0, 1, \ldots, N-1$ integers. The Fourier transformation of $\hat{H}^-$ reads

$$\hat{H}^- = \sum_{k \neq 0, \pi \in \mathcal{K}^-} (h - \cos k) \left(\hat{c}_k^\dagger \hat{c}_k + \hat{c}_{-k}^\dagger \hat{c}_{-k} \right) - \gamma \sin k \left(\hat{c}_k^\dagger \hat{c}_{-k}^\dagger + \hat{c}_{-k} \hat{c}_k \right)$$
$$+ (h-1) \left(\hat{c}_0^\dagger \hat{c}_0 - \hat{c}_0 \hat{c}_0^\dagger \right) + (h+1) \left(\hat{c}_\pi^\dagger \hat{c}_\pi - \hat{c}_\pi \hat{c}_\pi^\dagger \right), \tag{15}$$

where the 0-mode is occupied, whereas the π-mode is empty in the ground-state of $\hat{H}^-$. Performing a Bogoliubov rotation as in the case of $\hat{H}^+$, we obtain that the ground-state energy of $\hat{H}^-$ is given by

$$E_0^- = -\frac{1}{2} \sum_{k \in \mathcal{K}^-} s(k) \epsilon_k. \tag{16}$$

where

$$s(k) = \begin{cases} -1, & \text{if } k = 0, \\ 1, & \text{otherwise.} \end{cases} \tag{17}$$

Similarly to $\hat{H}^+$, the full spectrum of $\hat{H}^-$ is derived by adding an even number of fermions to the ground-state

$$E_k^- = E_0^- + \sum_{k \in \mathcal{P}_{\text{even}}(\mathcal{K}^-)} s(k) \epsilon_k, \tag{18}$$

where $\mathcal{P}_{\text{even}}(\mathcal{K}^-)$ is the subset of the power set of $\mathcal{K}^-$ with an even number of terms.

2.3 Gibbs Density Matrix

A system, with Hamiltonian $\hat{H}$, at thermal equilibrium with a bath at (inverse) temperature $\beta = \frac{1}{k_B T}$, where k_B is the Boltzmann constant, is described by a density matrix $\hat{\rho}(\beta, \hat{H})$, dubbed the Gibbs state: $\hat{\rho}(\beta, \hat{H}) = \frac{e^{-\beta \hat{H}}}{\mathcal{Z}(\beta, \hat{H})}$, where $\mathcal{Z}(\beta, \hat{\mathcal{H}}) = \text{Tr}\left\{ e^{-\beta \hat{H}} \right\}$ is the partition function.

Because of the direct sum structure of the Hamiltonian in Eq. (6), the thermal state of the spin system can also be cast in a direct sum form:

$$\hat{\rho}(\beta, \hat{H}) = \sum_{n=0}^{2^{N-1}-1} \left(\frac{e^{-\beta E_n^+}}{\mathcal{Z}(\beta, \hat{H})} \left|E_n^+\right\rangle\left\langle E_n^+\right| + \frac{e^{-\beta E_n^-}}{\mathcal{Z}(\beta, \hat{H})} \left|E_n^-\right\rangle\left\langle E_n^-\right| \right), \tag{19}$$

where $\left|E_n^{\pm}\right\rangle$ are the eigenstates of $\hat{H}^{\pm}$ and $p_n = \frac{e^{-\beta E_n^{\pm}}}{\mathcal{Z}(\beta,\hat{H})}$ are the Boltzmann weights giving the probability distribution function (PDF) of the Gibbs distribution. The latter is composed by

$$\{p_n\} = \left\{p_n^+\right\} \cup \left\{p_n^-\right\}. \tag{20}$$

From the non-interacting nature of the Hamiltonian and its set of symmetries, it is evident that there is a high degree of degeneracy in the Hamiltonian spectrum. In the following we will analytically determine how many distinct p_n's appear in Eq. (20), in order to minimize the number of necessary variational parameters in the PQC, capable of reproducing the Boltzmann distribution, and hence the Gibbs state.

2.4 Degeneracies in the Energy Spectrum

Let us start identifying the number of distinct energy levels in the positive parity subspace. Because of the translational symmetry, the single-particle energy spectrum is invariant under momentum inversion $k \to -k$, as evident from Eq. (12) being an even function of k: $\epsilon_k = \epsilon_{-k}$.

Clearly, the state with zero fermions, i.e., the ground-state, is non-degenerate. The next set of energies we analyze is that with two fermions added to the ground-state, that is $E_n^{(2)} = E_0^+ + \epsilon_k + \epsilon_q$, where $k \neq q \in \mathcal{K}^+$. The number of 2-fermions energy levels is given by the binomial $\binom{N}{2}$ of which $\frac{N}{2}$ energy levels are non-degenerate. This can be readily seen, as degenerate levels occur in the form

$$\epsilon_{-k} + \epsilon_{-q} = \epsilon_k + \epsilon_q = \epsilon_k + \epsilon_{-q} = \epsilon_{-k} + \epsilon_q, \tag{21}$$

with $k \neq q$. On the other hand, non-degenerate levels occur in the form

$$\epsilon_k + \epsilon_{-k}. \tag{22}$$

Hence, non-degenerate levels in the two-fermions excitation spectrum are given by how many distinct k's can be selected in Eq. (22). This number is obtained by observing that the momenta $k \in \mathcal{K}^+$ are symmetric around 0, and hence the number of ways to chose one momentum k out of $\frac{N}{2}$ momenta k is trivially given by $\frac{N}{2}$. Moreover, as Eq. (21) give rise to the same energy, the degeneracy is always four-fold in this particle sector. To summarise, out of the $\binom{N}{2}$ energy levels, there are $\binom{\frac{N}{2}}{2}$ 4-fold degenerate energy levels, with the remaining $\frac{N}{2}$ being non-degenerate.

Moving to the 4-fermion excited states, one can proceed in a similar way by identifying, out of the $\binom{N}{4}$ 4-fermion excitation energy levels, as non-degenerate levels being those with energy

$$\epsilon_{-k} + \epsilon_k + \epsilon_{-q} + \epsilon_q , \tag{23}$$

where $k \neq q$. Hence, the non-degenerate levels of the 4-fermion excited states are $\binom{\frac{N}{2}}{2}$, corresponding to the number of distinct sets (k, q) that can be selected from the available momenta in the positive subset of $\mathcal{K}^+$. The degrees of degeneracy in the 4-particles sector are $(1, 4, 16)$ occurring, respectively, $\left(\binom{\frac{N}{2}}{2}, \binom{\frac{N}{2}}{1}\binom{\frac{N}{2}-1}{2}, \binom{\frac{N}{2}}{4}\right)$ times.

Generalizing the above procedure to an arbitrary n-fermion energy subspace, for a chain of length N, one obtains that out of the $\binom{N}{n}$ n-fermion excitation energy levels

- the number of 1-fold degenerate levels is $\binom{\frac{N}{2}}{\frac{n}{2}}$
- the number of 4-fold degenerate levels is $\binom{\frac{N}{2}}{\frac{n}{2}-1}\binom{\frac{N}{2}-(\frac{n}{2}-1)}{2}$
- the number of 16-fold degenerate levels is $\binom{\frac{N}{2}}{\frac{n}{2}-2}\binom{\frac{N}{2}-(\frac{n}{2}-2)}{4}$
- the number of 64-fold degenerate levels is $\binom{\frac{N}{2}}{\frac{n}{2}-3}\binom{\frac{N}{2}-(\frac{n}{2}-3)}{6}$
- the number of 4^j-fold degenerate levels is $\binom{\frac{N}{2}}{\frac{n}{2}-j}\binom{\frac{N}{2}-(\frac{n}{2}-j)}{2j}$

Clearly, the maximum degree of degeneracy in the n-fermion energy subspace is $4^j \leq \binom{N}{n}$, with j being a non-negative integer. The procedure just described determines the degree of degeneracy and the number of energy levels in each degenerate subspace of $\hat{H}^+$ can be straightforwardly applied also to the negative parity subspace, which will not be reported here to avoid redundancy.

3 Preparing Probability Distributions on a Quantum Computer

In this Section we move to the main aim of this chapter: preparing a thermal state via a VQA. First we focus on preparing the Boltzmann distribution of the XY model entailed by Eq. (20).

Grover and Rudolph [17] first described a procedure to prepare log-concave probability density functions on a quantum computer. However, this approach was later proven that it does not achieve a quantum speed-up in general [21]. Modern approaches resort to VQAs [22–24], whereby parametrized ansätze are used to prepare the probability distribution on a quantum register: by varying the parameters and minimizing a relevant cost function, such as the relative entropy. Similar, in this vein, is the generative adversarial network (GAN), where the classical optimizer is replaced by a classical neural network, acting as a discriminator for the probability distribution [25–27].

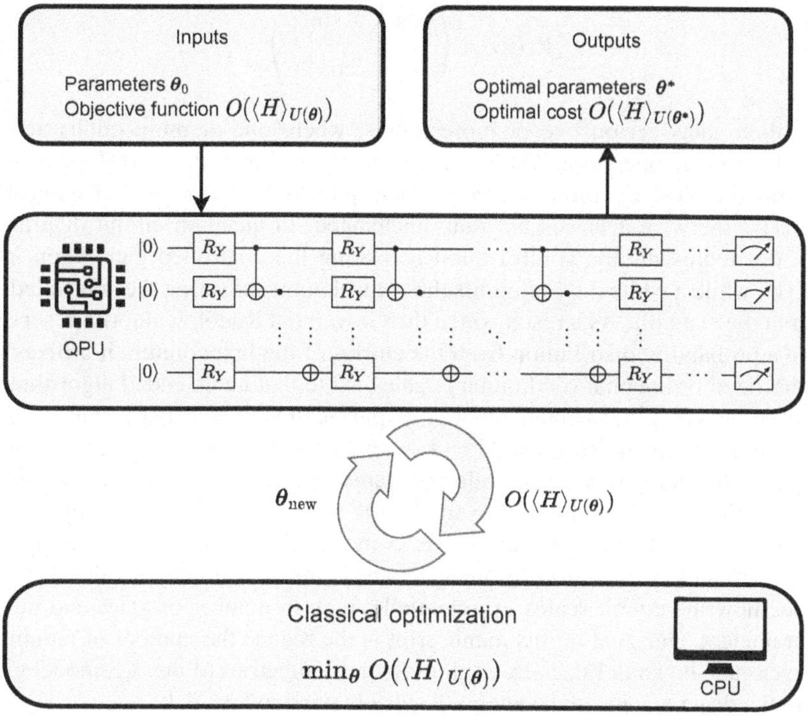

Fig. 2 Diagrammatic representation of a VQA. The framework of a VQA requires: an objective function (inputs), a parameterized quantum circuit (that operates on the quantum processing unit); and a classical optimizer (that operates on the classical processing unit). The Inputs are fed into the QPU (upper left arrow), the output of the PQC is fed into the CPU, which determines new parameters $\boldsymbol{\theta}_{\text{new}}$ provided to the QPU for the next iteration. This procedure is looped until optimal parameters $\boldsymbol{\theta}^*$ are found, which constitute the Outputs (upper right arrow)

A VQA comprises several modular components that can be readily combined, extended, and improved with developments in quantum hardware and theory [5]. These components include: the objective function, the cost function to be variationally optimized; the PQC, composed of unitary gates that are updated in the optimization procedure; and the classical optimizer, the method used to obtain the optimal circuit parameters that optimize the objective function. Figure 2 highlights the modules of a VQA.

To prepare the Boltzmann distribution of the XY model, we take the approach of utilizing a VQA, alongside a parametrized ansatz in the form of Grover and Rudolph [17]. In this case, an N-qubit ansatz requires $2^N - 1$ parameters, and $2^N - 1$ parameterized R_y gates, most of which are multi-controlled. An $R_y(\theta)$ gate acting on one qubit is given by

$$R_y(\theta) = \begin{pmatrix} \cos\frac{\theta}{2} & -\sin\frac{\theta}{2} \\ \sin\frac{\theta}{2} & \cos\frac{\theta}{2} \end{pmatrix}. \tag{24}$$

Controlled gates act on two or more qubits, where one or more qubits act as a control for some operation. The most common example is the CNOT gate, which performs the NOT operation on the second qubit only when the first qubit is the state $|1\rangle$, otherwise it leaves the state unchanged. In quantum circuit diagrams, a black dot represents the control qubit activating the controlled gate when in the state $|1\rangle$, while a white dot represents the control qubit activating the controlled gate when in the state $|0\rangle$. As a result, since the Grover and Rudolph algorithm serves to create a probability distribution from the entries of the first column, it utilizes only parametrized orthogonal (real unitary) gates. Note that an extended algorithm was proposed by Ref. [28] to obtain an optimal purification of an n-qudit state.

Let us consider the case for $N = 4$, with Fig. 3 showing the circuit. Each box represents the R_y gate with the relevant parameter, and the black and white dots represent the control qubits. For example, the gate with $R_y(\theta_9)$ acts only on qubit 3 if qubit 0 is in the state $|0\rangle$, qubit 1 is in the state $|1\rangle$, and qubit 2 is in the state $|0\rangle$. Since there are 2^k possible binary controls for a k-qubit controlled gate, one can see how the circuit scales exponentially, both in number of gates and number of parameters. Our goal in this manuscript is the reduce the number of parameters in the circuit shown in Fig. 3, by exploiting the symmetries of the XY model giving rise to the degeneracies in the energy levels obtained in Sec. 2.4.

The action of the circuit U in Fig. 3 acting on the all-zero four-qubit state results in

$$U\,|0000\rangle = \sum_{i=0000}^{1111} a_i\,|i\rangle\,. \tag{25}$$

The states $|0\rangle$ and $|1\rangle$ are the computational basis states of one qubit. As such, extending the computational basis states for n qubits is simply done by computing the tensor product of all combinations of the states $|0\rangle$ and $|1\rangle$ up to n qubits. For example, the computational basis states of two qubits are $|00\rangle\,, |01\rangle\,, |10\rangle\,, |11\rangle$. As a result, when measuring the quantum state in Eq. (25) in the computational basis, we end up with the following probabilities for each configuration:

$$\cos^2\left(\frac{\theta_0}{2}\right)\cos^2\left(\frac{\theta_1}{2}\right)\cos^2\left(\frac{\theta_3}{2}\right)\cos^2\left(\frac{\theta_7}{2}\right) = p_{0000}, \tag{26a}$$

$$\cos^2\left(\frac{\theta_0}{2}\right)\cos^2\left(\frac{\theta_1}{2}\right)\cos^2\left(\frac{\theta_3}{2}\right)\sin^2\left(\frac{\theta_7}{2}\right) = p_{0001}, \tag{26b}$$

$$\cos^2\left(\frac{\theta_0}{2}\right)\cos^2\left(\frac{\theta_1}{2}\right)\sin^2\left(\frac{\theta_3}{2}\right)\cos^2\left(\frac{\theta_8}{2}\right) = p_{0010}, \tag{26c}$$

$$\cos^2\left(\frac{\theta_0}{2}\right)\cos^2\left(\frac{\theta_1}{2}\right)\sin^2\left(\frac{\theta_3}{2}\right)\sin^2\left(\frac{\theta_8}{2}\right) = p_{0011}, \tag{26d}$$

$$\cos^2\left(\frac{\theta_0}{2}\right)\sin^2\left(\frac{\theta_1}{2}\right)\cos^2\left(\frac{\theta_4}{2}\right)\cos^2\left(\frac{\theta_9}{2}\right) = p_{0100}, \tag{26e}$$

$$\cos^2\left(\frac{\theta_0}{2}\right)\sin^2\left(\frac{\theta_1}{2}\right)\cos^2\left(\frac{\theta_4}{2}\right)\sin^2\left(\frac{\theta_9}{2}\right) = p_{0101}, \tag{26f}$$

$$\cos^2\left(\frac{\theta_0}{2}\right)\sin^2\left(\frac{\theta_1}{2}\right)\sin^2\left(\frac{\theta_4}{2}\right)\cos^2\left(\frac{\theta_{10}}{2}\right) = p_{0110}, \tag{26g}$$

$$\cos^2\left(\frac{\theta_0}{2}\right)\sin^2\left(\frac{\theta_1}{2}\right)\sin^2\left(\frac{\theta_4}{2}\right)\sin^2\left(\frac{\theta_{10}}{2}\right) = p_{0111}, \tag{26h}$$

$$\sin^2\left(\frac{\theta_0}{2}\right)\cos^2\left(\frac{\theta_2}{2}\right)\cos^2\left(\frac{\theta_5}{2}\right)\cos^2\left(\frac{\theta_{11}}{2}\right) = p_{1000}, \tag{26i}$$

$$\sin^2\left(\frac{\theta_0}{2}\right)\cos^2\left(\frac{\theta_2}{2}\right)\cos^2\left(\frac{\theta_5}{2}\right)\sin^2\left(\frac{\theta_{11}}{2}\right) = p_{1001}, \tag{26j}$$

$$\sin^2\left(\frac{\theta_0}{2}\right)\cos^2\left(\frac{\theta_2}{2}\right)\sin^2\left(\frac{\theta_5}{2}\right)\cos^2\left(\frac{\theta_{12}}{2}\right) = p_{1010}, \tag{26k}$$

$$\sin^2\left(\frac{\theta_0}{2}\right)\cos^2\left(\frac{\theta_2}{2}\right)\sin^2\left(\frac{\theta_5}{2}\right)\sin^2\left(\frac{\theta_{12}}{2}\right) = p_{1011}, \tag{26l}$$

$$\sin^2\left(\frac{\theta_0}{2}\right)\sin^2\left(\frac{\theta_2}{2}\right)\cos^2\left(\frac{\theta_6}{2}\right)\cos^2\left(\frac{\theta_{13}}{2}\right) = p_{1100}, \tag{26m}$$

$$\sin^2\left(\frac{\theta_0}{2}\right)\sin^2\left(\frac{\theta_2}{2}\right)\cos^2\left(\frac{\theta_6}{2}\right)\sin^2\left(\frac{\theta_{13}}{2}\right) = p_{1101}, \tag{26n}$$

$$\sin^2\left(\frac{\theta_0}{2}\right)\sin^2\left(\frac{\theta_2}{2}\right)\sin^2\left(\frac{\theta_6}{2}\right)\cos^2\left(\frac{\theta_{14}}{2}\right) = p_{1110}, \tag{26o}$$

$$\sin^2\left(\frac{\theta_0}{2}\right)\sin^2\left(\frac{\theta_2}{2}\right)\sin^2\left(\frac{\theta_6}{2}\right)\sin^2\left(\frac{\theta_{14}}{2}\right) = p_{1111}. \tag{26p}$$

If one knows beforehand the discrete probability distribution, then, by taking specific ratios of expressions in Eqs. (26), one can relate the parameters θ_i to ratios of probabilities, e.g.,

$$\frac{p_{1111}}{p_{1110}} = \tan^2\left(\frac{\theta_{14}}{2}\right) \implies \theta_{14} = 2\arctan\left(\sqrt{\frac{p_{1111}}{p_{1110}}}\right). \tag{27}$$

By iterating this procedure for all θ, one can readily obtain

$$\theta_{l+2^{N-k}-1} = 2\arctan\left(\sqrt{\frac{\sum_{i=0}^{k-1} p_{2^k l+i+k}}{\sum_{j=0}^{k-1} p_{2^k l+j}}}\right), \tag{28}$$

where k ranges from $1, \ldots, N$, and l from $0, \ldots, 2^{N-k}-1$.

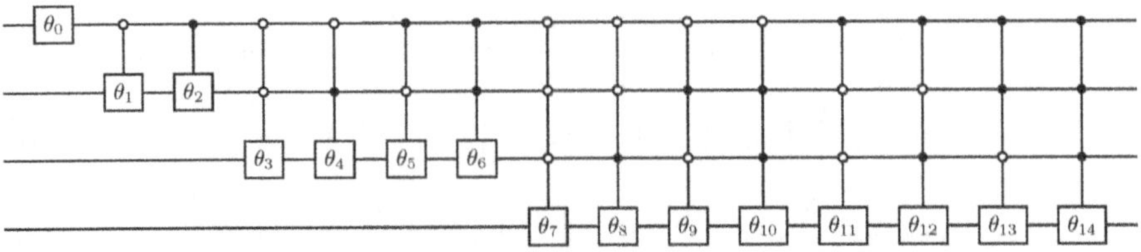

Fig. 3 Grover and Rudolph Circuit for preparing an arbitrary probability distribution embedded in a real quantum state. Each gate is represented an R_y gate with its respective parameter θ_i. Black dots represent the control qubit activating on $|1\rangle$, while the white dots represent the control qubit activating on $|0\rangle$

4 Few-Body *XY* chain

To illustrate an instance of the Gibbs PDF preparation via the Grover and Rudolph algorithm where the degeneracies of the energy levels allow for a reduction of the number of variational parameters needed in the PQC, let us restrict to a few-body scenario. This allows to better keep track of the protocol relying on the analytic expressions of the eigenenergies. Let us focus on the $N = 4$ XY chain.

Since the XY model is non-interacting, the spectrum is given by the single-particle energies, as discussed in Sect. 2. The model is also symmetric, when the eigenenergies are calculated with the modes ordered lexicographically, we can assign the first half of the computational basis states to cover $\{p_n^+\}$, with the latter half to cover $\{p_n^-\}$. The eigenenergies are:

$$E_0^+ = \frac{-\sqrt{\gamma^2 + 2h\left(h - \sqrt{2}\right) + 1} - \sqrt{\gamma^2 + 2h\left(h + \sqrt{2}\right) + 1}}{\sqrt{2}}, \tag{29a}$$

$$E_1^+ = \frac{\sqrt{\gamma^2 + 2h\left(h - \sqrt{2}\right) + 1} - \sqrt{\gamma^2 + 2h\left(h + \sqrt{2}\right) + 1}}{\sqrt{2}}, \tag{29b}$$

$$E_2^+ = E_3^+ = E_4^+ = E_5^+ = 0, \tag{29c}$$

$$E_6^+ = \frac{-\sqrt{\gamma^2 + 2h\left(h - \sqrt{2}\right) + 1} + \sqrt{\gamma^2 + 2h\left(h + \sqrt{2}\right) + 1}}{\sqrt{2}}, \tag{29d}$$

$$E_7^+ = \frac{\sqrt{\gamma^2 + 2h\left(h - \sqrt{2}\right) + 1} + \sqrt{\gamma^2 + 2h\left(h + \sqrt{2}\right) + 1}}{\sqrt{2}}, \tag{29e}$$

$$E_0^- = -1 - \sqrt{\gamma^2 + h^2},\ E_1^- = 1 - \sqrt{\gamma^2 + h^2}, \tag{29f}$$

$$E_2^- = E_3^- = -E_4^- = -E_5^- = -h, \tag{29g}$$

$$E_6^- = -1 + \sqrt{\gamma^2 + h^2},\ E_7^- = 1 + \sqrt{\gamma^2 + h^2}. \tag{29h}$$

We start to look at the Boltzmann coefficients generated by the eigenenergies in the negative parity sector. In the case of the ratio

$$\frac{p_{1111}}{p_{1110}} = \tan^2\left(\frac{\theta_{14}}{2}\right) = e^{-\beta(E_7^- - E_6^-)} = e^{-2\beta} \implies \theta_{14} = 2\arctan\left(e^{-\beta}\right). \tag{30}$$

Next we find that

$$\frac{p_{1101}}{p_{1100}} = \tan^2\left(\frac{\theta_{13}}{2}\right) = e^{-\beta(E_5^- - E_4^-)} = 1 \implies \theta_{13} = \frac{\pi}{2}, \tag{31}$$

and

$$\frac{p_{1011}}{p_{1010}} = \tan^2\left(\frac{\theta_{12}}{2}\right) = e^{-\beta(E_3^- - E_2^-)} = 1 \implies \theta_{12} = \frac{\pi}{2}, \tag{32}$$

due to the pairs of degeneracies. Furthermore

$$\frac{p_{1001}}{p_{1000}} = \tan^2\left(\frac{\theta_{11}}{2}\right) = e^{-\beta(E_1^- - E_0^-)} = e^{-2\beta} \implies \theta_{11} = 2\arctan\left(e^{-\beta}\right). \tag{33}$$

Now looking at the Boltzmann coefficients given by the eigenenergies in the positive parity sector:

$$\frac{p_{0101}}{p_{0100}} = \tan^2\left(\frac{\theta_9}{2}\right) = e^{-\beta(E_5^+ - E_4^+)} = 1 \implies \theta_9 = \frac{\pi}{2}, \tag{34}$$

and

$$\frac{p_{0011}}{p_{0010}} = \tan^2\left(\frac{\theta_8}{2}\right) = e^{-\beta(E_3^+ - E_2^+)} = 1 \implies \theta_8 = \frac{\pi}{2}. \tag{35}$$

Moreover, while we cannot determine the exact value of θ_{10} and θ_7, due to their dependency on γ and h, we can still relate them through

$$\frac{p_{0111}}{p_{0110}} = \frac{p_{0001}}{p_{0000}} \implies \tan^2\left(\frac{\theta_{10}}{2}\right) = \tan^2\left(\frac{\theta_7}{2}\right) \implies \theta_{10} = \theta_7. \tag{36}$$

Finally, we also have that

$$\frac{p_{0101}}{p_{0011}} = \frac{\cos^2\left(\frac{\theta_4}{2}\right)}{\sin^2\left(\frac{\theta_3}{2}\right)}\tan^2\left(\frac{\theta_1}{2}\right) = e^{-\beta(E_5^+ - E_3^+)} = 1$$
$$\implies \theta_4 = 2\arccos\left(\frac{\sin\left(\frac{\theta_3}{2}\right)}{\tan\left(\frac{\theta_1}{2}\right)}\right). \tag{37}$$

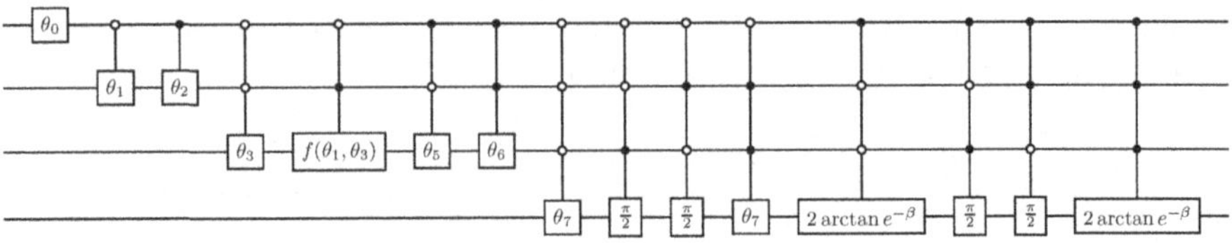

Fig. 4 Figure 3 with substituted parameters, Eqs. (30)–(37), where $f(\theta_1, \theta_3) = 2 \arccos\left(\frac{\sin\left(\frac{\theta_3}{2}\right)}{\tan\left(\frac{\theta_1}{2}\right)}\right)$

Thus we have managed to eliminate 8 parameters from a total of 15. This may be due to the fact that we have N single-particle energy levels in both the positive and the negative parity subspace. However we also have the fact that in the negative parity subspace, the sum of the energies of the π- and 0-modes sum always up to 2, hence giving an additional constraint. This result may hint towards the fact that for the XY model composed of N particles, the number of parameters needed to characterize its Boltzmann distribution is $2N - 1$. Figure 4 represents the PQC in Fig. 3 with the reduced number of parameters obtained from Eqs. (30)–(37).

5 Structure of the VQA

We utilize the algorithm proposed in Ref. [16] to prepare Gibbs states of the XY model. Suppose one chooses a Hamiltonian $\hat{H}$ and inverse temperature β. For a general state $\hat{\rho}$, one can define a generalized Helmholtz free energy as

$$\mathcal{F}(\hat{\rho}) = \mathrm{Tr}\left\{\hat{H}\hat{\rho}\right\} - \beta^{-1}\mathcal{S}(\hat{\rho}), \tag{38}$$

where the von Neumann entropy $\mathcal{S}(\hat{\rho})$ can be expressed in terms of the eigenvalues, p_i, of $\hat{\rho}$,

$$\mathcal{S}(\hat{\rho}) = -\sum_{i=0}^{d-1} p_i \ln p_i. \tag{39}$$

Since the Gibbs state is the unique state that minimizes the free energy, a variational form can be put forward that takes Eq. (38) as an objective function, such that

$$\hat{\rho}(\beta, \hat{H}) = \arg\min_{\hat{\rho}} \mathcal{F}(\hat{\rho}). \tag{40}$$

In this case, $p_i = \exp(-\beta E_i) / \mathcal{Z}(\beta, \hat{H})$ is the probability of getting the eigenstate $|E_i\rangle$ from the ensemble $\hat{\rho}(\beta, \hat{H})$.

The PQC, as shown in Fig. 5, is composed of a unitary gate U_A acting on the ancillary qubits, and a unitary gate U_S acting on the system qubits, with CNOT gates

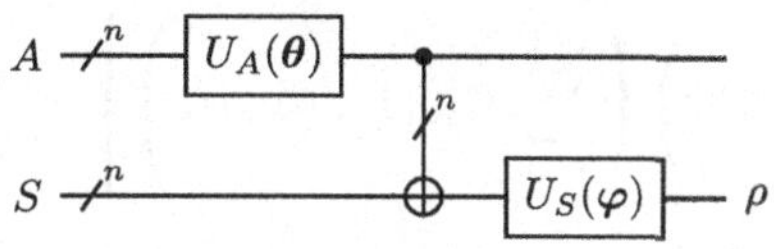

Fig. 5 PQC for Gibbs state preparation, with systems A and S each carrying N qubits. CNOT gates act between each qubit A_i and corresponding S_i

in between. Note that the circuit notation we are using here means that there are N qubits for both the system and the ancillae, as well as N CNOT gates that act in parallel, $\text{CNOT}_{AS} \equiv \bigotimes_{i=0}^{N-1} \text{CNOT}_{A_i S_i}$. The parametrized unitary U_A acting on the ancillae, followed by CNOT gates between the ancillary and system qubits, is responsible for preparing a probability distribution on the system. The parametrized unitary U_S is then applied on the system qubits to transform the computational basis states into the eigenstates of the Hamiltonian.

Throughout the paper we will describe unitary operations and density matrices in the computational basis. We will also denote the N-fold tensor product of a state $|\psi\rangle$, of an N-qubit register R, as $|\psi\rangle_R^{\otimes N} \equiv \bigotimes_{i=0}^{N-1} |\psi\rangle_i$, where $i \in R$. Denote a general unitary gate of dimension $d = 2^N$, as $U_A = (u_{i,j})_{0 \leq i,j \leq d-1}$. Starting with the initial state of the $2n$-qubit register, $|0\rangle_{AS}^{\otimes 2n}$, we apply the unitary gate U_A on the ancillae to get a quantum state $|\psi\rangle_A$, such that $(U_A \otimes \hat{\mathbb{1}}_S) |0\rangle_{AS}^{\otimes 2n} = |\psi\rangle_A \otimes |0\rangle_S^{\otimes N}$, where $|\psi\rangle_A = \sum_{i=0}^{d-1} u_{i,0} |i\rangle_A$ and $\hat{\mathbb{1}}_S$ is the identity acting on the system. Since U_A is applied to the all-zero state in the ancillary register, then the operation serves to extract the first column of the unitary operator U_A, as discussed in Sect. 3. The next step is to prepare a classical probability mixture on the system qubits, which can be done by applying CNOT gates between each ancilla and system qubit. This results in a state

$$\text{CNOT}_{AS} \left(|\psi\rangle_A \otimes |0\rangle_S^{\otimes N} \right) = \sum_{i=0}^{d-1} u_{i,0} |i\rangle_A \otimes |i\rangle_S . \tag{41}$$

By then tracing out the ancillary qubits, we arrive at

$$\text{Tr}_A \left\{ \left(\sum_{i=0}^{d-1} u_{i,0} |i\rangle_A \otimes |i\rangle_S \right) \left(\sum_{j=0}^{d-1} u_{j,0}^* \langle j|_A \otimes \langle j|_S \right) \right\} = \sum_{i=0}^{d-1} |u_{i,0}|^2 |i\rangle \langle i|_S , \tag{42}$$

ending up with a diagonal mixed state on the system, with probabilities given directly by the absolute square of the entries of the first column of U_A, that is, $p_i = |u_{i,0}|^2$. If the system qubits were traced out instead, we would end up with the same diagonal mixed state,

$$\text{Tr}_S \left\{ \left(\sum_{i=0}^{d-1} u_{i,0} |i\rangle_A \otimes |i\rangle_S \right) \left(\sum_{j=0}^{d-1} u_{j,0}^* \langle j|_A \otimes \langle j|_S \right) \right\} = \sum_{i=0}^{d-1} |u_{i,0}|^2 |i\rangle \langle i|_A . \quad (43)$$

This implies that by measuring in the computational basis of the ancillary qubits, we can determine the probabilities p_i, which can then be post-processed to determine the von Neumann entropy $\mathcal{S}$ of the state $\hat{\rho}$ via Eq. (39) (since the entropy of A is the same as that of S).

The unitary gate U_S then serves to transform the computational basis states of the system qubits to the eigenstates of the Gibbs state, once it is optimized by the VQA, such that

$$\hat{\rho} = U_S \left(\sum_{i=0}^{d-1} |u_{i,0}|^2 |i\rangle \langle i|_S \right) U_S^\dagger = \sum_{i=0}^{d-1} p_i |\psi_i\rangle \langle \psi_i| , \quad (44)$$

where the expectation value $\text{Tr}\left\{\hat{H}\hat{\rho}\right\}$ of the Hamiltonian can be measured. Ideally, at the end of the optimization procedure, $p_i = \exp(-\beta E_i) / \mathcal{Z}(\beta, \hat{H})$ and $|\psi_i\rangle = |E_i\rangle$, so that we get

$$\hat{\rho}(\beta, \hat{H}) = \sum_{i=0}^{d-1} \frac{e^{-\beta E_i}}{\mathcal{Z}(\beta, \hat{H})} |E_i\rangle \langle E_i| . \quad (45)$$

Finally, we define the objective function of our VQA to minimize the free energy (38), via our constructed PQC, to obtain the Gibbs state

$$\begin{aligned} \hat{\rho}(\beta, \hat{H}) &= \arg\min_{\boldsymbol{\theta}, \boldsymbol{\varphi}} \mathcal{F}\left(\hat{\rho}\left(\boldsymbol{\theta}, \boldsymbol{\varphi}\right)\right) \\ &= \arg\min_{\boldsymbol{\theta}, \boldsymbol{\varphi}} \left(\text{Tr}\left\{ \hat{H} \hat{\rho}_S(\boldsymbol{\theta}, \boldsymbol{\varphi}) \right\} - \beta^{-1} \mathcal{S}\left(\hat{\rho}_A(\boldsymbol{\theta})\right) \right) . \end{aligned} \quad (46)$$

After optimization, the system register will (ideally) contain the prepared Gibbs state. In Refs. [16, 29], U_A is the ansatz responsible for preparing the Boltzmann distribution, while U_S is the ansatz responsible for preparing the eigenstates of the Hamiltonian. In this work, we replace U_A in Fig. 5 by the ansatz in Fig. 4, and optimize to obtain the Gibbs state of the XY model. U_S is a parity-preserving ansatz, given by a brick-wall layering of R_P gates, denoted as

$$R_P(\varphi_i, \varphi_j) = \begin{pmatrix} \cos\left(\frac{\varphi_i+\varphi_j}{2}\right) & 0 & 0 & \sin\left(\frac{\varphi_i+\varphi_j}{2}\right) \\ 0 & \cos\left(\frac{\varphi_i-\varphi_j}{2}\right) & -\sin\left(\frac{\varphi_i-\varphi_j}{2}\right) & 0 \\ 0 & \sin\left(\frac{\varphi_i-\varphi_j}{2}\right) & \cos\left(\frac{\varphi_i-\varphi_j}{2}\right) & 0 \\ -\sin\left(\frac{\varphi_i+\varphi_j}{2}\right) & 0 & 0 & \cos\left(\frac{\varphi_i+\varphi_j}{2}\right) \end{pmatrix} . \quad (47)$$

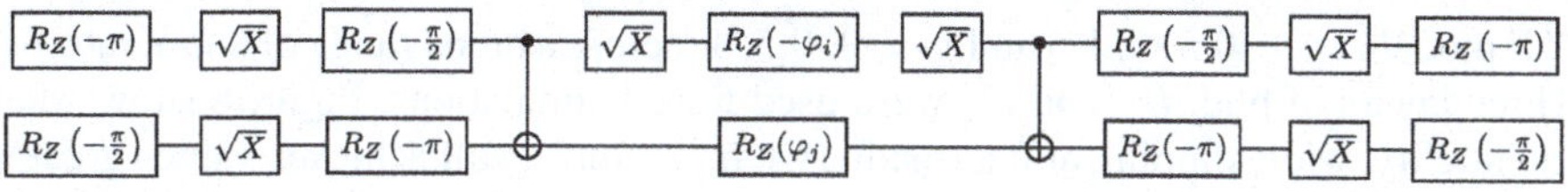

Fig. 6 Decomposed R_P gate in Eq. (47) in IBM native basis gates

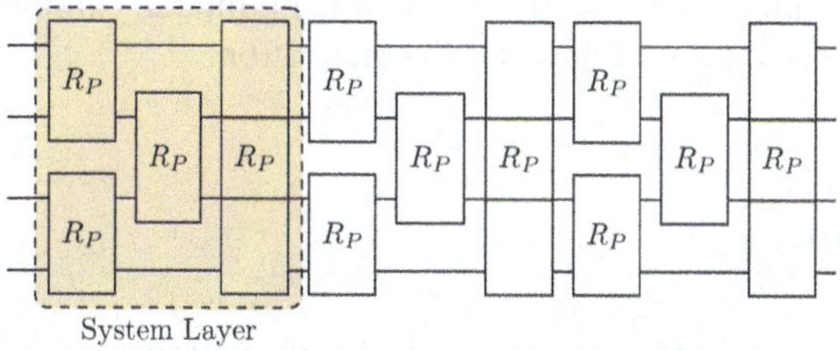

Fig. 7 A representation of the PQC embodying U_S tasked with performing the rotation form the computational to the eigenenergy basis of the system Hamiltonian. The shaded area represent a single brick-wall layer that is repeated three times in obtaining the fidelity of Fig. 8. Note that each R_P gate contains 2 parameters as in Eq. (47)

As one can see from the unitary itself, any input state will have the same parity as the output state, and so, an ansatz constructed solely of parity-preserving gates will also be parity-preserving. The decomposed unitary gate is shown in Fig. 6, while the entire circuit of U_S is shown in Fig. 7.

To assess the performance of the VQA, we utilize the Uhlmann-Josza fidelity as a figure of merit [30], defined as $F\left(\hat{\rho}, \hat{\sigma}\right) = \left(\mathrm{Tr}\left\{\sqrt{\sqrt{\hat{\rho}}\hat{\sigma}\sqrt{\hat{\rho}}}\right\}\right)^2$. This fidelity measure quantifies the "closeness" between the prepared state and the target Gibbs state, making it a commonly-employed metric for distinguishability. Figure 8 shows the fidelity of the prepared Gibbs state with the theoretical Gibbs state of the four-qubit

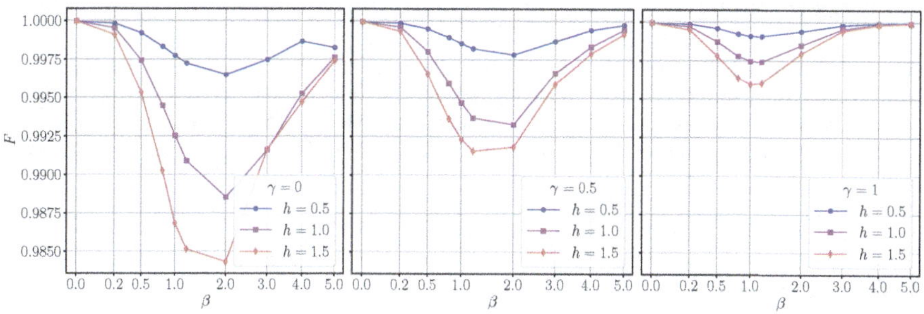

Fig. 8 Fidelity F between the obtained statevector simulations (using BFGS) and the exact Gibbs state, versus inverse temperature β, for the XY model with $h = 0.5, 1, 1.5$ and $\gamma = 0, 0.5, 1$ in Eq. (2) for four qubits. A total of 100 runs are made for each point, with the optimal state taken to be the one that maximizes the fidelity

XY model at $h = 0.5, 1, 1.5$ and $\gamma = 0, 0.5, 1$, across a broad range of temperatures. Three layers of both U_A and U_S were used for all simulations. Figure 8 shows that the fidelity, while dipping at intermediary temperatures, still achieves values greater than 98% for all models and temperatures. Interestingly, the results for $\gamma = 0$ appear to be slightly less optimal. This may due to the fact that the XX model belongs to a different universality class than the XY and Ising model, exhibiting an additional symmetry, the conservation of the total magnetization.

6 Conclusion

In this chapter we have addressed the problem of preparing a quantum many-body thermal state via a variational quantum algorithm suitable for NISQ computers. We utilized a parametrized quantum circuit composed of two blocks: the former acting on an ancillary register tasked with reproducing the Boltzmann distribution, with the latter acting on the system register tasked with rotating the system's state from the computational basis to the eigenenergy basis. The two registers are connected by a set of CNOT gates, and the optimizer is tasked with minimizing the generalized Helmholtz energy in order to prepare the Gibbs state on the system register.

We have improved on the Grover and Rudolph procedure for the preparation of an arbitrary probability distribution function, by reducing the number of exponential parameters needed in the VQA exploiting the symmetries of the XY spin-$\frac{1}{2}$ Heisenberg model. We have then applied our protocol to a four-site chain and showed that our VQA is able to reach almost unit fidelity in statevector simulations by combining the Grover and Rudolph algorithm, with a reduced number of variational parameters acting on the ancillary register, and a parity-preserving ansatz on the system register. Although we have applied our protocol to small system sizes, our results hint toward the fact that a sub-exponential number of variational parameters are necessary for the preparation of a many-body Gibbs distribution by exploiting the degeneracies in the energy spectrum.

Acknowledgements The Authors acknowledge discussions with Francesco Plasina and Jacopo Settino on the topic of this chapter. TJGA ackowledges funding through the IPAS+ (Internationalisation Partnership Awards Scheme +) QIM project by the MCST (The Malta Council for Science & Technology). M.C. acknowledges funding by TESS (Tertiary Education Scholarships Scheme).

References

1. Bengtsson, I., Zyczkowski, K.: Geometry of Quantum States: An Introduction to Quantum Entanglement. Cambridge University Press, Cambridge (2006). https://doi.org/10.1017/cbo9780511535048
2. Barahona, F.: On the computational complexity of Ising spin glass models. J. Phys. A: Math. Gen. **15**(10), 3241–3253 (1982). https://doi.org/10.1088/0305-4470/15/10/028

3. Harrow, A.W., Montanaro, A.: Quantum computational supremacy. Nature **549**(7671), 203–209 (2017). https://doi.org/10.1038/nature23458. (arXiv:1809.07442)
4. Preskill, J.: Quantum Computing in the NISQ era and beyond. Quantum **2**, 79 (2018). https://doi.org/10.22331/q-2018-08-06-79
5. Bharti, K., Cervera-Lierta, A., Kyaw, T.H., Haug, T., Alperin-Lea, S., Anand, A., Degroote, M., Heimonen, H., Kottmann, J.S., Menke, T., Mok, W.-K., Sim, S., Kwek, L.-C., Aspuru-Guzik, A.: Noisy intermediate-scale quantum algorithms. Rev. Mod. Phys. **94**(1), 015004 (2022). https://doi.org/10.1103/RevModPhys.94.015004
6. Cerezo, M., Arrasmith, A., Babbush, R., Benjamin, S.C., Endo, S., Fujii, K., McClean, J.R., Mitarai, K., Yuan, X., Cincio, L., Coles, P.J.: Variational quantum algorithms. Nat. Rev. Phys. **3**(9), 625–644 (2021). https://doi.org/10.1038/s42254-021-00348-9. arxiv:2012.09265
7. Tilly, J., Chen, H., Cao, S., Picozzi, D., Setia, K., Li, Y., Grant, E., Wossnig, L., Rungger, I., Booth, G.H., Tennyson, J.: The variational quantum eigensolver: a review of methods and best practices. Phys. Rep. **986**, 1–128 (2022). https://doi.org/10.1016/j.physrep.2022.08.003. (arXiv:2111.05176)
8. Ratini, L., Capecci, C., Benfenati, F., Guidoni, L.: Wave function adapted Hamiltonians for quantum computing. J. Chem. Theory Comput. **18**(2), 899–909 (2022). https://doi.org/10.1021/acs.jctc.1c01170
9. Consiglio, M., Chetcuti, W.J., Bravo-Prieto, C., Ramos-Calderer, S., Minguzzi, A., Latorre, J.I., Amico, L., Apollaro, T.J.G.: Variational quantum eigensolver for SU(N) fermions. J. Phys. A: Math. Theor. **55**(26), 265301 (2022). https://doi.org/10.1088/1751-8121/ac7016. (arXiv:2106.15552)
10. Consiglio, M., Apollaro, T.J.G., Wieśniak, M.: Variational approach to the quantum separability problem. Phys. Rev. A **106**(6), 062413 (2022). https://doi.org/10.1103/PhysRevA.106.062413
11. Watrous, J.: Quantum Computational Complexity (2008). arXiv:0804.3401
12. Holmes, Z., Muraleedharan, G., Somma, R.D., Subasi, Y., Şahinoğlu, B.: Quantum algorithms from fluctuation theorems: thermal-state preparation. Quantum **6**, 825 (2022). https://doi.org/10.22331/q-2022-10-06-825
13. Motta, M., Sun, C., Tan, A.T.K., O'Rourke, M.J., Ye, E., Minnich, A.J., Fernando, G.S.L., Brandão, F.G.S.L., Chan, G.K.-L.: Determining eigenstates and thermal states on a quantum computer using quantum imaginary time evolution. Nat. Phys. **16**(2), 205–210 (2020). https://doi.org/10.1038/s41567-019-0704-4
14. Haug, T., Bharti, K.: Generalized quantum assisted simulator. Quantum Sci. Technol. **7**(4), 045019 (2022). https://doi.org/10.1088/2058-9565/ac83e7
15. Schaller, G.: Adiabatic preparation without quantum phase transitions. Phys. Rev. A **78**(3), 032328 (2008). https://doi.org/10.1103/PhysRevA.78.032328
16. Consiglio, M., Settino, J., Giordano, A., Mastroianni, C., Plastina, F., Lorenzo, S., Maniscalco, S., Goold, J., Apollaro, T.J.G.: Variational Gibbs state preparation on NISQ devices (2023). arXiv:2303.11276
17. Grover, L., Rudolph, T.: Creating superpositions that correspond to efficiently integrable probability distributions (2002). https://arxiv.org/abs/quant-ph/0208112
18. Lieb, E., Schultz, T., Mattis, D.: Two soluble models of an antiferromagnetic chain. Ann. Phys. **16**(3), 407–466 (1961). https://doi.org/10.1016/0003-4916(61)90115-4
19. Damski, B., Rams, M.M.: Exact results for fidelity susceptibility of the quantum Ising model: the interplay between parity, system size, and magnetic field. J. Phys. A: Math. Theor. **47**(2), 025303 (2014). https://doi.org/10.1088/1751-8113/47/2/025303
20. Franchini, F.: An introduction to integrable techniques for one-dimensional quantum systems. In: Lecture Notes in Physics, vol. 940. Springer International Publishing, Cham (2017). https://doi.org/10.1007/978-3-319-48487-7. arXiv:1609.02100
21. Herbert, S.: No quantum speedup with Grover-Rudolph state preparation for quantum monte Carlo integration. Phys. Rev. E **103**, 063302 (2021). https://doi.org/10.1103/PhysRevE.103.063302
22. Nakaji, K., Uno, S., Suzuki, Y., Raymond, R., Onodera, T., Tanaka, T., Tezuka, H., Mitsuda, N., Yamamoto, N.: Approximate amplitude encoding in shallow parameterized quantum circuits

and its application to financial market indicators. Phys. Rev. Res. **4**(2), 023136 (2022). https://doi.org/10.1103/PhysRevResearch.4.023136
23. Dasgupta, K., Paine, B.: Loading probability distributions in a quantum circuit (2022). arXiv:2208.13372
24. Kumar, R.T., Raina, A.: Generating probability distributions using variational quantum circuits (2023). arXiv:2307.09147
25. Zoufal, C., Lucchi, A., Woerner, S.: Quantum generative adversarial networks for learning and loading random distributions. npj Quantum Inf. **5**(1), 103 (2019). https://doi.org/10.1038/s41534-019-0223-2
26. Situ, H., He, Z., Wang, Y., Li, L., Zheng, S.: Quantum generative adversarial network for generating discrete distribution. Inf. Sci. **538**, 193–208 (2020). https://doi.org/10.1016/j.ins.2020.05.127
27. Agliardi, G., Prati, E.: Optimal tuning of quantum generative adversarial networks for multivariate distribution loading. Quantum Rep. **4**(1), 75–105 (2022). https://doi.org/10.3390/quantum4010006
28. Benenti, G., Strini, G.: Optimal purification of a generic n-qudit state. Phys. Rev. A **79**(5), 052301 (2009). https://doi.org/10.1103/PhysRevA.79.052301
29. Consiglio, M.: Variational quantum algorithms for Gibbs state preparation (2023). arXiv:2305.17713
30. Uhlmann, A.: Transition probability (fidelity) and its relatives. Found. Phys. **41**(3), 288–298 (2011). https://doi.org/10.1007/s10701-009-9381-y. (arXiv:1106.0979)

Implementing a Quantum Information Engine Using Spintronics

Martin Bowen and Mathieu Lamblin

Abstract Quantum information engines can transform information gained from a quantum measurement into available work. Beyond theoretical constructs, such engines are typically studied using optical/microwave strokes on localized electron states (e.g. NV^- centers in diamond). This Chapter describes a spin electronic implementation of this engine construct. After reviewing prior spintronic experiments that might be described as manifestations of this engine class, we present recent experiments and theory that describe how a quantum information engine can be assembled using spintronics. Herein, we emphasize the importance of the ferromagnetic metal/molecule interface as the apparatus to measure the quantum spin state of fluctuating paramagnetic centers borne by stand-alone and molecule-embedded atoms. This construct utilizes the underpinnings of its solid-state design to advantageously exhibit inherent quantum coherence between these paramagnetic atoms, along a quantization axis defined by the measurement apparatus. We highlight the need for further insight using auxiliary experiments. Retooling the industrial chain that commercializes magnetic memories to implement this concept could enable mid-term applications to harvest ambient thermal energy.

1 Introduction

A quantum information engine is a cyclical set of processes by which the measurement of quantum information on discrete electronic states forming the engine's working substance procures additional free energy that can be used to perform work [1–14]. Although mechanical [15] and ion [16] or particle [17] trap implementations exist, proposals on engine cycles that use this quantum resource typically involve optical strokes to craft and measure the discrete quantum state that constitute the engine's working substance (WS). Only a few experiments [2, 18, 19] on optical implementations have been reported so far.

M. Bowen (✉) · M. Lamblin
Institut de Physique et Chimie des Matériaux de Strasbourg, UMR 7504, CNRS Université de Strasbourg, 23 rue du Loess BP 43, 67034 Strasbourg Cedex, France
e-mail: martin.bowen@ipcms.unistra.fr

X. Bouju and C. Joachim (eds.), *Crossroad of Maxwell Demon*, Advances in Atom and Single Molecule Machines, https://doi.org/10.1007/978-3-031-57904-2_5

This historical focus on optical quantum information engines (QIEs) underscores several de facto advantages as the light-matter interaction can (1) lead to precise changes onto a WS's discrete electronic state, so that precise concepts can be tested; (2) be tailored to control sources of quantum decoherence; and (3) address an ensemble of WS states that are independent from one another. This final point advantageously enables large aggregate readout signals from what is originally a single atom light-matter interaction. This strategy also means that the WS electronic states that are otherwise isolated from the outside world can be used at high/room temperature. The most widely used WS here is the NV^- center in diamond [2, 19]. As a major disadvantage for possible applications, optical QIEs require substantial auxiliary equipment: laser/microwave sources, optical benches, etc.... This reliance also extends to possible feedback mechanisms acting on the WS quantum state. This makes implementing on-chip solutions technically challenging, and industrial applications difficult. And it also means that interactions are externally turned on/off, yet their energy cost is not accounted for theoretically.

A purely solid-state electronic QIE implementation could in principle lift these caveats, in particular if the electronic implementation of the strokes is inherently achieved through built-in electronic interactions between the engine's solid-state components [20–28]. However, for almost all implementations, several of the aforementioned advantages are lost. The main difficulty is to craft both the WS and electrical leads to it so as to fulfill the criteria of a WS with few, discrete, quantum entangled states that is connected to the outside world over a carefully tailored quantum channel. On one side, scanning tunneling spectroscopy experiments operate on single atoms/molecules, and could in principle be well suited to the task by crafting the electronic transmission between the target atom and the substrate/tip [26]. However, such experiments are so far lacking [28] and require auxiliary equipment. Another approach is to utilize semiconductor quantum dots [29], which are used in single electron transistors. But the failure of such an approach to deliver a quantum advantage hinges on a terminology-induced misconception: single electron transistors may operate at the single electron level, but in fact involve many electrons! Indeed, the electronic puddle created by electrostatically confining a 2D electron gas typically contains ~100 electrons. Device operation then occurs between the 100th and 101st levels. The resulting low energy difference not only confines device operation to sub-Kelvin temperatures, but the mesoscopic WS cannot sustain the quantum coherence needed to operate the engine using quantum information. As a result, mesoscopic QD-based electronic engines do not exhibit quantum-based efficiency enhancements [29]. A more promising, well-identified track is to use superconducting qubits [27], though here as well, the superconducting requirement to obtain quantum coherence and control over electronic transport also carries a stringent temperature requirement.

The present chapter discusses recent advances into a spintronic implementation of a QIE that is autonomous, on-chip/self-contained and operates at/above room temperature. Spintronics is a green nanotechnology with industrial penetration in which the electron spin is used to encode and transmit information [30]. Energy applications of spintronics have, aside from the material in this chapter, been essentially confined to classical thermoelectrics [31] and to the harvesting of artificial microwave energy

[32]. To establish the milestones of discovery and rationalization, we adopt a historical approach by first presenting early results [24, 33, 34] whose reports lacked a QIE backdrop in Sect. 2. We then lay out concepts, theory [35] and a rationalized experiment [25] by the Bowen group on the spintronic QIE in Sect. 3. The final Sect. 4 will provide conclusions and outline links between the various research communities, from magnetization dynamics to atom-level electronic physicists [28], that can actively contribute to accelerate research on the topic.

2 First Results

The research field of spintronics originated in the magnetic thin film community with the discovery by Nobel prize winners Peter Grünberg and Albert Fert of interlayer exchange coupling and giant magnetoresistance (MR) across ferromagnetic/non-magnetic metallic multilayers [36]. Attempts to increase the MR device performance metric led to modifications to the spacer layer and magnetic electrode materials (Sect. 2.1). In the process of studying these materials modifications, three groups published results within 2009–2019 on what appears, retrospectively, to be implementations of a quantum spintronic engine (Sect. 2.2).

2.1 *Materials Context*

We first discuss changes to spacer materials (Sect. 2.1), and then to electrode materials (Sect. 2.2), noting that in some instances, the two may combine for enhanced spintronic performance.

2.1.1 Spacer Layer Materials

Magnetic tunnel junctions, featuring first amorphous AlOx and then almost exclusively epitaxial/polycrystalline MgO, have enhanced the metal-only device magnetoresistance (MR) metric of spintronic performance with tunnel magnetoresistance *TMR* values as high as 600% at room temperature [37]. The Julliere model [38] relates this metric to the effective spin polarization P of the current flowing from/to the ferromagnetic metallic electrodes:

$$TMR = \frac{2P_L P_R}{1 - P_L P_R},$$

where $P_L(P_R)$ refers to the polarization of the left (right) electrode. MgO-based MTJs achieve high performance through a synergy between the band structures of the *fcc*(001)-oriented MgO tunnel and of the *bcc*-alloy Fe electrodes that straddle the barrier [30]. This materials track has generated considerable industrial uptake with applications ranging from magnetic memories and neuromorphic computing [39] to harvesters of waste microwave energy [32]. MgO-based magnetic memories in particular have achieved industrial penetration within several companies: it is this class of spintronic devices that epitomizes the commercial potential of spintronic technologies.

Other strategies have involved using ferromagnetic insulators as the spacer layer. Here, the spin-split bands generate different barrier heights for spin up and down carriers, such that a moderately spin-polarized current from a ferromagnetic metallic electrode can be further spin filtered upon tunneling through such a barrier material. Examples include EuO and EuS [40].

2.1.2 Electrode Materials

3d transition metals exhibit a spin polarization *P* of transport electronic state of 40–50%. One strategy to increase *P* is to utilize a so-called half-metallic material. This material is characterized by the effective presence at the Fermi surface of transport electrons of only one spin sub-band. This property, which usually arises from electron correlations, can be found in complex oxides such as $La_{0.7}Sr_{0.3}MnO_3$ [41], but was also predicted [42, 43] to occur in dilute magnetic semiconductors. Much research in the 2000s thus focused on integrating 3d atoms within III-V (e.g. GaAs:Mn; MnAs) and II-VI (ZnO:Co) semiconductors. With a view to transport electron spin over large distances, this strategy also intended to overcome resistivity mismatch issues [44] when injecting spin-polarized electrons into the semiconductor by integrating the spin-polarized transport property within a semiconducting matrix.

Another route, discovered in the 2010s, consists in decorating the surface of *3d* transition metal with molecules. The interface state that results from *p-d* bonds was found to exhibit a high spin polarization of electronic states at the Fermi level according to spectroscopy measurements [45]. In fact, the very experiments on the quantum spintronic engine (QSE) that are described in later sections demonstrated that electronic transport across this so-called 'spinterface' is also highly spin-polarized [24, 25], with effective values of $MR = 770\%$ corresponding to $P = 89\%$. There is a debate as to whether this is a generic effect [45] or reflects specific aspects of energy level alignments [46] between the molecule's orbitals and the electrode states. We will further describe properties of the spinterface when describing the QSE.

2.2 *Experiments 2009–1019*

We now describe the three experiments that historically report observations of what, in retrospect, are thought to be manifestations of a QSE. All experiments were conducted with the nominal aim of improving spintronic performance, using barrier and electrode materials described in the previous section. Any deviations from the present QSE model, detailed in the following Section, are mentioned.

In 2009, the Barnes and Maekawa groups reported on magnetotransport experiments at 10 K across MnAs nanoparticles within a GaAs matrix [33]. Although one MnAs ferromagnetic (FM) electrode was used (Fig. 1a), large values of MR were observed. This was attributed to spin-polarized carriers on the MnAs nanoparticles that act as the analyzer of spin-polarized transport originating from the MnAs electrode. Interestingly, when at low temperature, applying an external magnetic field H past a certain critical value causes the IV to shift into the $(V > 0, I < 0)$ power generating quadrant (see Fig. 1b). Removing H caused a return to the passive device state. This switchover from a passive to an active device state is confirmed using $I(H)$ curves (Fig. 1c). An enduring spontaneous bias voltage was observed over minutes, even when a passive resistance load was connected.

In 2014, the Moodera group reported on magnetotransport experiments at 1 K across Al nanoparticles sandwiched between EuS spin filter tunnel barriers [34] (see Fig. 2a). The authors observed a spontaneous bias voltage across the device whose amplitude and sign depend on the relative orientation of the two ferromagnetic tunnel barriers' magnetizations (Fig. 2b). This spintronic manipulation also leads to a change in the spontaneous current (see Fig. 2c) upon connecting the device to a load resistance, which stays constant up to ~10 MΩ load. This implies an output power of 27 fW at 1 K.

In 2019, the Bowen group reported on magnetotransport experiments at 295 K across C atoms within a MgO magnetic tunnel junction (MTJ) with Co/C spinterfaces (Fig. 3). Spintronic interactions (Fig. 3a) cause the observation of up to 0.1 nW of output power and peculiar features such as spectroscopic data with sub-$k_B T$ spectral resolution (Fig. 3b). An analytical model involving a C dimer sandwiched between fully spin-polarized quantum dots (the spinterfaces; Fig. 3c) is used to reproduce the experimental data (Fig. 3d). The model supposes the presence of a spin-split chemical potential on the leads (i.e. a phenomenological accounting of the work performed), evidences the role of spintronic anisotropy, i.e. the change in magnetic anisotropy of a junction's components due to a spin-polarized current [47], and describes the net current flow as the result of an imbalance in oppositely flowing spin channels of charge current. The Bowen group report implements room temperature operation and high spintronic performance across the industrial vector of MgO MTJs [48]. The main weakness of the work was that the effect was observed on only one device out of 216. This underscores the difficulty to control the insertion and position of individual impurities within an ultrathin oxide layer.

We now briefly discuss how the effects observed were at the time explained. The Barnes and Maekawa groups' report attributed the observed spontaneous bias voltage to the spin motive force [33], a higher-order correction to Faraday's law. The explanation in terms of the magnetic energy of paramagnetic clusters was criticized at the time [49]. Referring to the present model of the QSE (see Section III), this implementation lacks two spin-polarized electrodes, and the hexagonal MnAs thin film is not expected to have a high spin-polarization. The Moodera [34] and Bowen [24] groups' articles attributed the power output to harvesting spin fluctuations on

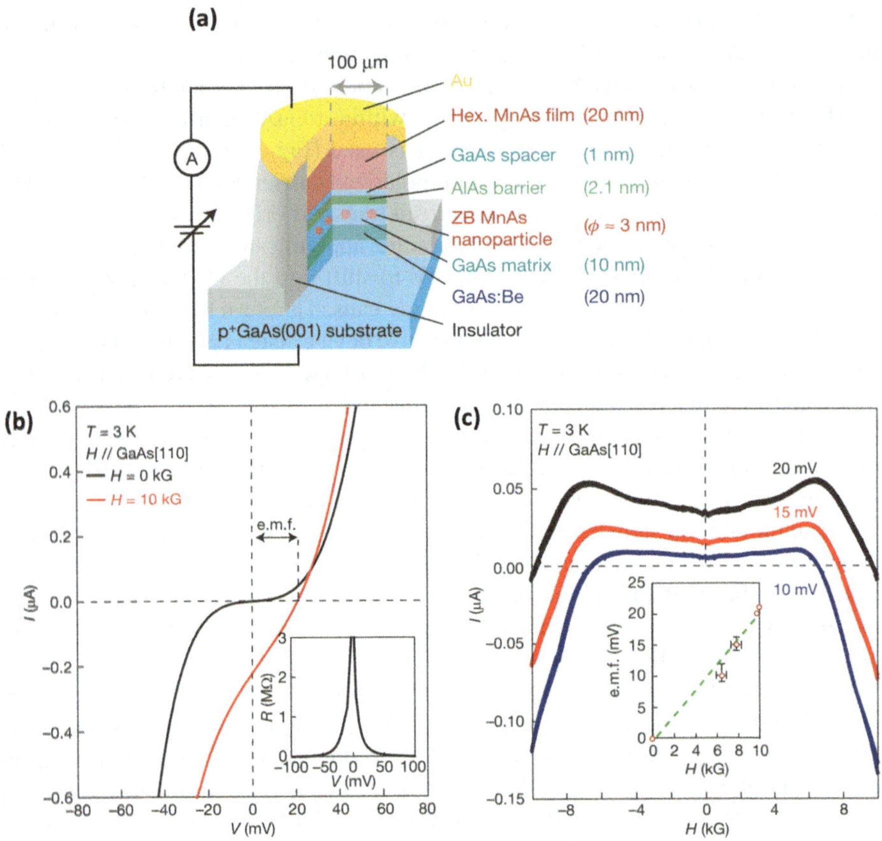

Fig. 1 The **Barnes and Maekawa groups' report in 2009**. **a** Device schematic, showing the single MnAs ferromagnetic electrode and MnAs nanoparticles. **b** I(V) data for H = 0 and H = 10 kOe. The magnetic field induces current and bias offsets that are confirmed in **c** I(H) data. This indicates a spintronic origin of the effect. Adapted with the permission from Ref. [33] Copyright 2009 Nature

paramagnetic centers using spin-polarized currents that generate spin-splitting on the PM centers and pin the electrode Fermi level to the corresponding spin state. It was in turn proposed, without thermodynamics proof, that the ensuing potential profile drop leads to an effective potential bias between the electrodes. This Fermi level pinning to a spin state would require a highly spin-polarized current. The Moodera report [34] utilizes ferromagnetic insulators as the source of spin-polarized current, yet there is no experimental evidence that this spin polarization is very high. The Bowen report [24] utilizes the ferromagnetic metal/molecule interface with a purportedly high spin polarization according to spin-polarized photoemission, and experimentally find a 79% spin polarization in transport experiments across the Co/C interface—a record at the time.

Assuming this potential drop, the Bowen group proposed [24] a phenomenological model of transport (see above), and described basic electronic interactions that result

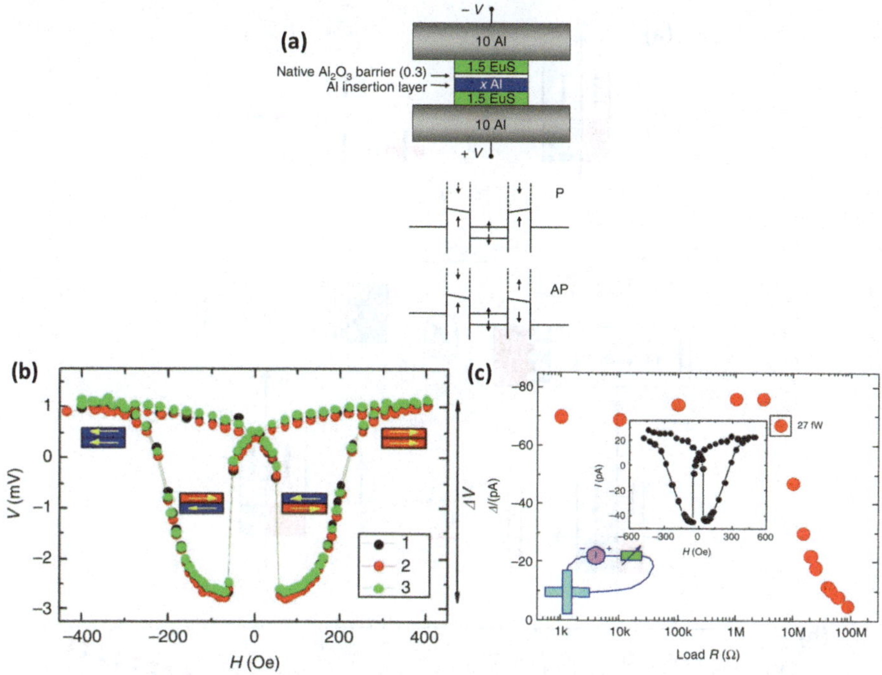

Fig. 2 The **Moodera group report in 2014**. **a** Schematics of the device stack and of the effective potential landscape in the junction's two parallel (P) and antiparallel (AP) magnetic configurations, which are proposed to depend on the pinning of the Fermi level to the nanoparticle states. **b** Spontaneous bias voltage versus applied magnetic field. **c** Change in current due to P $\leftrightarrow$ AP as a function of load resistance amplitude. Experiments were performed at T = 1 K. Adapted with the permission from Ref. [34] Copyright 2014 Nature

from the spinterface—paramagnetic center interaction, cooling on the PM center and perturbations to the electrode ferromagnetic ground state. When describing the needed intersection between spintronics and quantum thermodynamics to explain the effect, the Bowen group report alluded to quantum thermal [22, 23, 50] engines, but the link with quantum information/coherence engines remained tenuous [51]. Nevertheless, the Bowen group report includes sub-$k_B T$ spectral features, which is presently thought to be an experimental signature of the QSE feedback mechanism (see Sect. 3.2).

3 The Quantum Spintronic Engine: Present Understanding

We begin this section by presenting an overview of the quantum thermodynamics concepts that the Bowen group invoked in a 2022 report [25] to explain observations of power generation across several molecular spintronic junctions. This

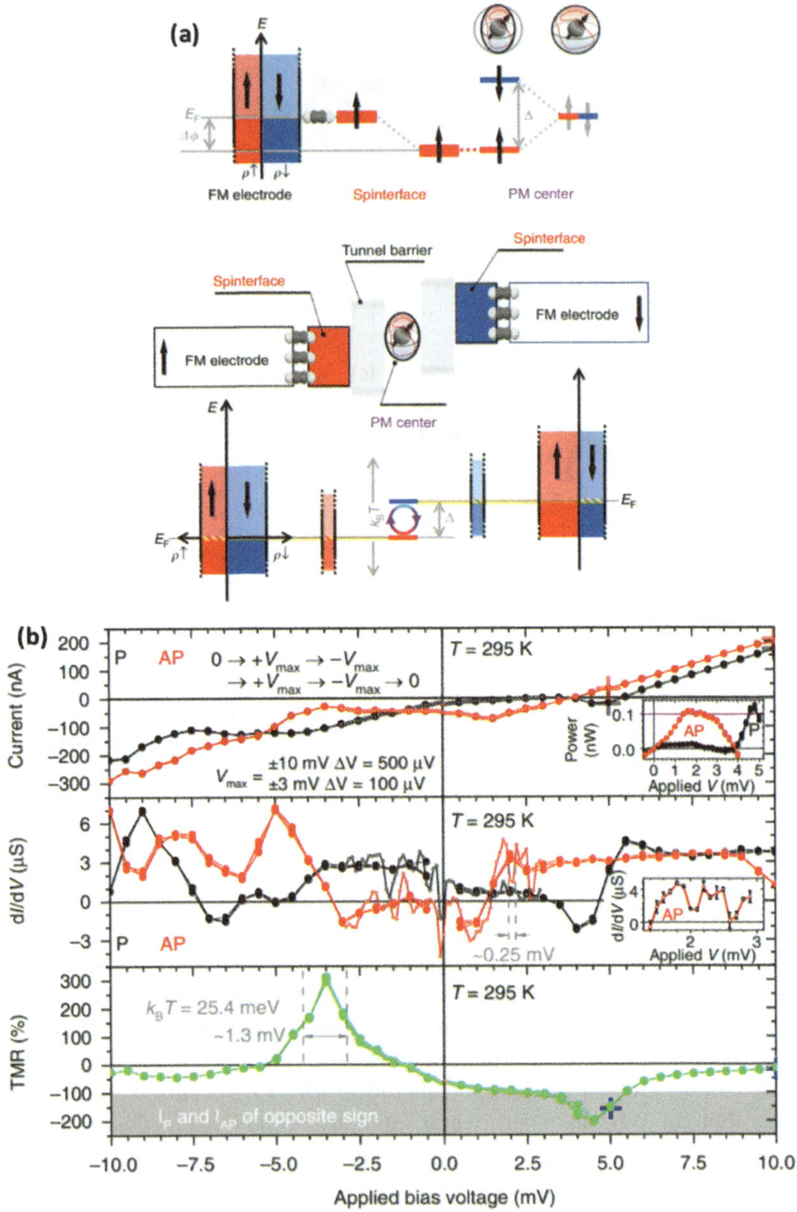

Fig. 3 The **Bowen group report in 2019.** **a** Schematics of the electronic potential landscape due to the effects of spin-polarized charge transfer, including Fermi level pinning and a spin-splitting on the PM center. **b** Magnetotransport experiments at 295 K across a magnetic tunnel junction with Co/C interfaces and likely C atoms in the MgO barrier. The data reveal power generation, sub-k_BT spectral features, and large TMR. **c** Schematic of the analytical model to phenomenologically reproduce, **d** experimental data. The model describes the impact of spintronic anisotropy on the effective spin splitting, and junction transport as originating from two oppositely flowing spin channels of current. Adapted with the permission from Ref. [24] Copyright 2019 Nature

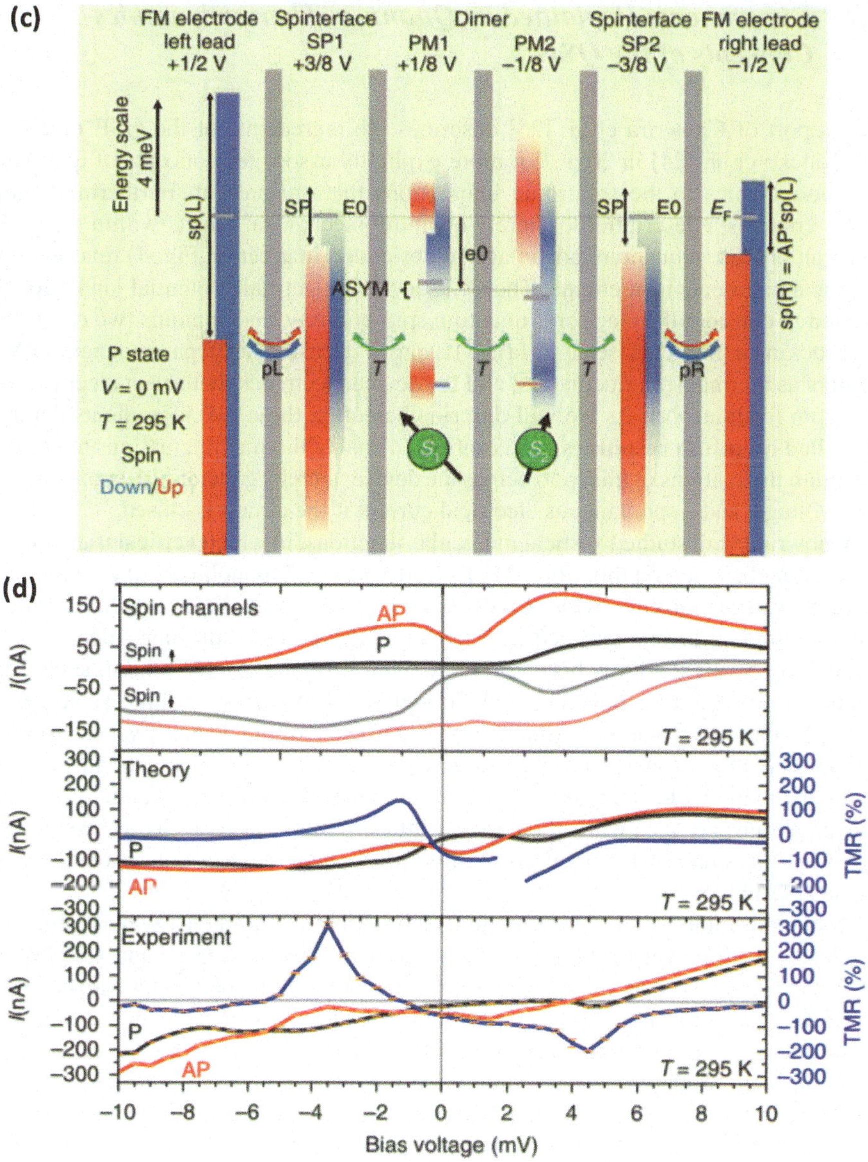

Fig. 3 (continued)

report strongly bridges, for the first time, the spintronics and quantum thermodynamics communities. In a second part, we describe the main features of a standalone atomic theory [35] that explains how to harvest quantum fluctuations and thermal fluctuations on paramagnetic centers when connected to fully spin-polarized leads.

3.1 Experiments Explained by Quantum Thermodynamics Concepts of the QSE

The report of Chowrira et al. [25] describes the ingredients of the QSE much as did Katcko et al. [24] in 2019, but more explicitly associates concepts of quantum thermodynamics to the spintronic interactions that are present. Furthermore, the paper contains a description of purely quantum mechanical effects within the QSE that can procure a quantum advantage (color-coded in green in Fig. 4) relative to a classical nanoscale heat engine. The QSE device's electronic potential landscape is crafted to consider the electron's quantum spin property, and contains two carefully interlocking ingredients (see Fig. 4a): (1) a single, or ensemble of, paramagnetic (PM) centers as thermal spin fluctuators; and (2) device electrodes that select electrons of one spin for transport. As we will describe hereafter, these two ingredients deploy so-called **quantum resources** and combine to enable thermal fluctuations to rectify quantum fluctuations of transport across the device, thereby generating a spontaneous bias voltage, and a spontaneous electrical current if the circuit is closed.

Chowrira et al. studied vertical molecular junctions, in which vertical transport is proceeding between Fe thin film FM electrodes across 3 monolayers of Co phthalocyanine (CoPc) molecules with a Co PM center. The Fe and CoPc layers are separated by ultrathin C_{60} layers, and Fe/C_{60} interfaces [45] are the spintronic selector. The fluctuating $S = 1/2$ spin is borne by the Co out-of-plane dz^2 [2] orbital, such that spins can couple antiferromagnetically through super-exchange, forming a spin chain [21]. Despite the mesoscopic (diameter 300 nm) size of the device's vertical pillar, transport is also expected to proceed across a reduced number of nano-transport paths involving CoPc spin chains. Indeed, the intra-chain superexchange promotes the main transport path, while there is very little crosstalk between chains due to very low wavefunction overlap between neighboring CoPc molecules within a molecular plane [21, 53, 55–57].

Using this molecular implementation, Chowrira et al. report energy harvesting on 9 devices, with power reaching 24 nW at 360 K, 27 nW at 295 K, and 450 nW at 40 K (see Fig. 4e), i.e. a 270 × increase over the previous record set by Katcko et al. [24]. According to Chowrira et al. [25], the crucial point is to implement so-called quantum resources that place the engine's WS and thermal reservoirs out of thermodynamic equilibrium [58]. This applies if the WS exhibits a coherent superposition of quantum states with population inversion [2] or a phase transition [54]. The engine's reservoirs can be 'non-thermal' if the thermal stroke (i) has an action that is faster than the thermalization or decoherence processes ('Zeno effect'), (ii) that exhibits some asymmetry between the two quadratures of conjugated quantities ('squeezed bath'[52]) or (iii) if it injects quantum coherence into the WS [1]. Chowrira et al. argue that the experimental data can be interpreted as bearing the signature of two quantum resources: squeezed baths and a WS at a phase transition [54]. Indeed, referring to Fig. 4a, a spectral resolution 100 × better than k_BT is observed (i.e. similar to the 2019 report [24]), despite the usual 2–3k_BT broadening expected during electronic transport. This is characteristic of a feedback-driven transport [59–62] that

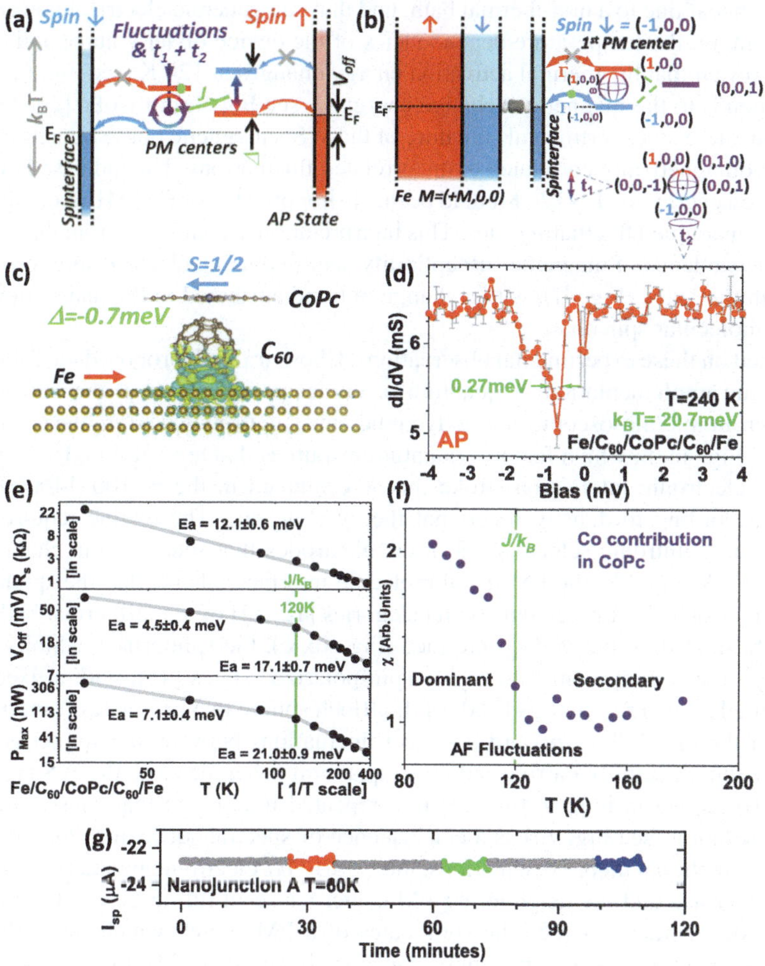

Fig. 4 A molecular spintronic quantum engine: concept and experimental signatures of quantum resources (green color-coded in caption/panels a–f). a Illustration of the WS and spinterfaces [45, 46] interacting due to quantum-correlated strokes. The PM centers are magnetically coupled with energy J. The spinterface is an example of a spintronic selector. **b** Illustration of the heat stroke between the Fe/C_{60} spinterface, with fixed (−1,0,0) spin polarization relative to the Fe static magnetization, and the Co PM center of CoPc that is spin-split due to coherent (−1,0,0) spin injection. Spin mixing on the PM center due to decoherence (t_1/t_2), iSWAP and transport effects are shown in both panels. **c** Density functional theory calculations of the corresponding Fe/C_{60}/CoPc system. C_{60} ↔ CoPc charge transfer isn't visible on this color scale. (d) dI/dV across devices with CoPc PM centers reveals sub-$k_B T$ spectral features, i.e. a signature of the non-thermal [1, 52] bath due to the spinterface-mediated heat stroke. **e** Thermal activation plots of R, V_{Off} and P reveal a transition at 120 K between two regimes for V_{Off} and P, corresponding [21, 53] to J (panel a) for CoPc [21, 53]. **f** Confirmation of this interpretation from the temperature dependence of magnetic susceptibility, acquired using Electron Spin Resonance (ESR) on a similar stack. This is a signature within engine performance of the phase transition [54] of the CoPc WS. **g** Time dependence of the spontaneous current across a CoPc device. Current/potential offsets are order of magnitude beyond those found using calibrated resistances. Adapted with the permission from Ref. [25] Copyright 2022, Wiley

would appear due to a non-thermal bath, and that can generate electrical power [63]. Furthermore, the temperature dependencies of the device offset voltage and output power reveal that the thermal activation energy changes at 120 K (see Fig. 4e). This corresponds to the magnetic exchange energy (J, see Fig. 4a) of CoPc [21, 53], i.e. can constitute an experimental signature of the WS's magnetic phase transition from antiferromagnetically correlated to uncorrelated fluctuations. The increased thermal activation energy for $T > 120$ K could be due to hopping transport [64] across the spin chain's uncoupled fluctuating sites. This interpretation is confirmed from the temperature dependence of magnetic susceptibility acquired using electron spin resonance on a similar stack (Fig. 4f), with a change in trend attributed to the phase transition of the molecular spin chain.

Based on these experimental observations, Chowrira et al. proposed the following spintronic implementation of quantum resources in an engine at the intersection between atomic/mesoscopic thermodynamics and spintronics. The engine is schematized in Fig. 4a, b, which features quantum resources that are color-coded in green, and an electronic interaction stroke that is evaluated in the $\sim$ 100–140000 GHz range according to density functional theory (Fig. 4c). The engine requires at a minimum 'spintronic selectors' (SS), i.e. electrodes that select a spin channel for transport. As one SS, the FM metal/molecule interface, also called the spinterface [46], spintronically implements both categories [12, 52] of a non-thermal bath with the WS. Indeed, at the Fe/C_{60} interface (Fig. 4b, c), the spinterface exhibits a low density of spectrally confined, highly spin-polarized states [45], with a fixed spin referential along $(x,y,z) = (-1,0,0)$ of the Bloch sphere that originates from the FM state of the metal. The forward transport fluctuations between the spinterface and WS are thus spectrally narrow and inject quantum coherence into the WS [1] along $(-1,0,0)$ (coded in blue in Fig. 4b) in a repeated interaction that causes the sub-k_BT resolution (see Fig. 4d) as a consequence of spectral squeezing. According to density functional theory calculations, the $\sim$140 THz electronic interaction between the spinterface and the neighboring PM center causes a lifting $\Delta = -0.7$ meV of energy degeneracy between the spin states of a PM center, which fits within the 1.4 meV experimental energy window (see Fig. 1c and Ref. [24]) that was observed. Note how the mesoscopic (diameter 300 nm) device's effective nano-transport path [65] should comprise several PM centers forming a spin chain, across which small energy asymmetries can be present [24].

According to Chowrira et al., these properties not only describe both categories of non-thermal baths, but also underscore the quantum coupling between strokes in this continuous regime engine [2]. Indeed, by lifting the spin degeneracy on the WS (i.e. spintronic anisotropy [47, 57, 66]), the spinterface-based stroke promotes a superposition of quantum states of the WS upon which spin mixing can promote population inversion. This generation of quantum ressources occurs due to transport [24, 66, 67] and quantum iSWAP [68] processes moderated by J, while spin fluctuations introduce both $(\pm 1,0,0)$ spin relaxation and orthogonal $(0,, \pm 1, \pm 1)$ phase decoherence (coded in purple in Fig. 1b). With $k_BT > |\Delta| = 0.7$ meV, spin fluctuations are operative up to 169.5 GHz. Since the 1 MHz phase decoherence frequency of CoPc [69] is much slower than the 140 THz spinterface stroke and the 169.5 GHz spin mixing stroke, the engine is resilient against phase decoherence.

The reverse transport fluctuation to the spinterface can only occur from the (−1,0,0) level of the WS, so that any electron transfer from the WS to the spinterface requires a return to the ground state, i.e. a quantum resource against which work can be performed. This could explain the extremely long spontaneous currents (stable over 2 h) that are observed (Fig. 4g).

To support the claim of device electrodes with a high spin polarization, Chowrira et al. performed magneto-transport measurements (see Fig. 5). In panel (a), the raw I(H) data show that, depending on the value of applied bias, the device can block current in either the P or AP magnetic state, leading to magneto-current values of ~1500%. The current direction can even be reserved by switching the magnetic state. I(V) data shown in panel (b) reveal the presence of different bias offsets for each junction magnetic state. Finally, in panel (c), the slope resistance is plotted as a function of external applied magnetic field, and reveals a magnetoresistance MR = 770% that corresponds to a record spin polarization P = 89% at 40 K in of the Fe/C_{60} spinterface (see Fig. 5).

3.2 *Formal Quantum Thermodynamical Theory of the QSE*

The previous subsection presented Chowrira et al.'s description in broad terms of how a QSE may be expected to operate by superimposing concepts that are present in literature on the thermodynamics of quantum information onto experimental data, and given the nominal structure of the junctions studied. To place this description on more solid footing, Lamblin and Bowen reported a quantum thermodynamics theory [35] on a two-stroke engine that can harvest some of the energy of quantum fluctuations and thermal fluctuations on spin sites into a directed electrical current.

Figure 6a displays a schematic of the QSE energy levels and interactions proposed by Lamblin and Bowen. Mimicking the proposed description by Chowrira et al., the model describes a WS composed of two spin sites that are magnetically coupled with energy J to one another, and with an on-site Coulombic repulsion term U. All energy levels are serially coupled by spin-conserving tunnel interactions (γ parameters) such that the WS is connected to fully spin-polarized leads that represent the spinterface and electrode in experiments [24, 25]. To simplify the problem and obtain an analytical solution, one sublevel on the right spin is placed at higher energy outside the transport/electronic interaction window. The physical justification is that, as Chowrira used a thinner C_{60} spacer on one side of junction ($\gamma_R >> \gamma_L$), the effective magnetic field acting on the center should lead to a much larger spin splitting. The other essential simplification, based on the prior report, is antiferromagnetic coupling between spin sites ($J > 0$). Lastly, for simplicity, spintronic anisotropy was assumed to be constant, implying a constant bias voltage [24] so that the final Hamiltonian of the system reads [35]:

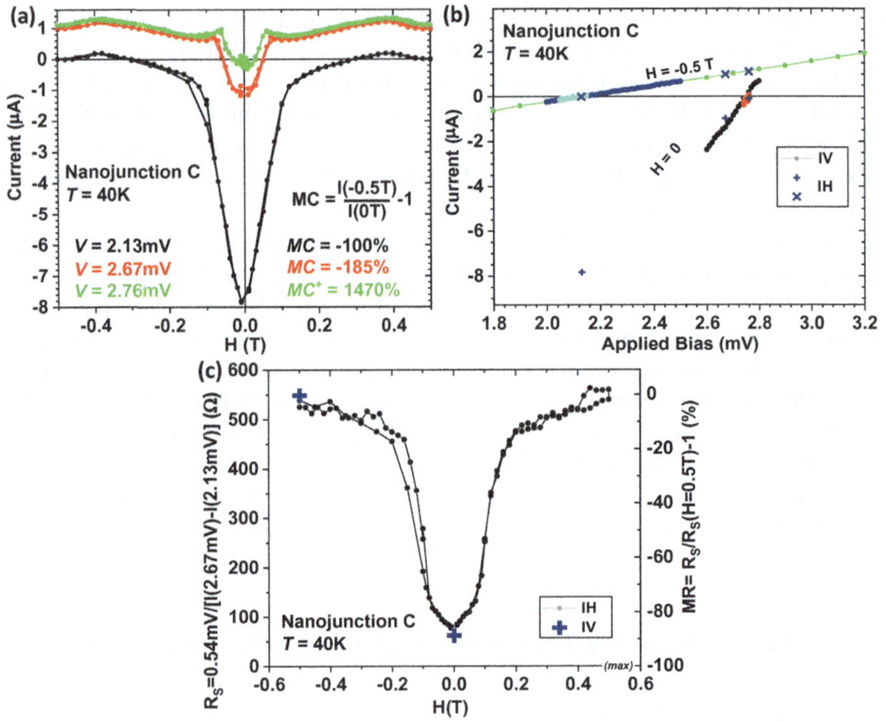

Fig. 5 Spintronic features of the CoPc molecular thermal energy harvester. **a** I(H) data. **b** I(V) data at H = 0 T and H = −0.5 T, revealing a linear behavior with $R_S = 63\ \Omega$ and $R_S = 550\ \Omega$ around $V_{Off} = 2.13$ mV and $V_{Off} = 2.76$ mV, respectively. The blue crosses reflect the I(H) data from panel a. **c** R_S(H) calculated from two I(H) datasets from panel (**a**). The blue crosses indicate R_S inferred from panel **b**. The two spintronic V_{Off} lead to extremal values of $-100 < \text{MC}\ (\%) < \infty$ in I(H) data, while R_S indicates MR = −79% (or optimistic MR = + 77%, i.e. P = 89%). Current/bias offsets are order of magnitude beyond those found using calibrated resistances [25]

$$H_S = \epsilon_{L\uparrow} n_{L\uparrow} + \epsilon_{L\downarrow} n_{L\downarrow} + \epsilon_R n_R + \gamma c^\dagger_{L\downarrow} c_R + \gamma^* c^\dagger_R c_{L\downarrow} + J n_{L\downarrow} n_R + U n_{L\uparrow} n_{L\downarrow}.$$

Here $c^\dagger_R$ and c_R are the creation/annihilation operators of the down spin on the right side, i.e. the spin site with stronger on-site spin splitting. On the left-hand spin site, spin up and spin down annihilation operators are denoted $c_\uparrow$ and $c_\downarrow$ and ϵ are the on-site effective energies. From the illustration and the associated Hamiltonian, we can thus see that only one transport path connects the two electrodes across the spin dimer, but this path requires a spin-flip to be operative. The Hamiltonian is thus designed to test the role of quantum fluctuations as the only source of spin-flip events, unless otherwise noted. Naturally, this also entails that these spin sites be described by atoms, rather than a mesoscopic quantum dot containing ∼ 100 electrons.

Lamblin and Bowen describe the QSE as containing features that are specific to the spintronic experimental underpinnings. Indeed, since electronic interactions are

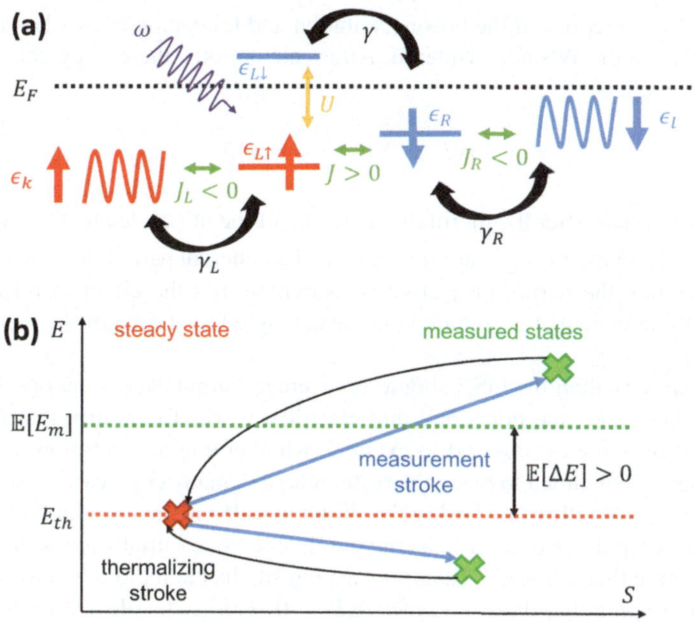

Fig. 6 First formal theory of the Quantum Spintronic Engine: design. a Illustration of the potential levels of two atomic spin quantum dots connected to fully spin-polarized leads. The right spin site is strongly coupled to the lead while the left site's spin occupation can vary depending on quantum fluctuations and, optionally, bosonic catalysis (ω). **b** Energy-Entropy diagram of the engine's two measurement and thermalization strokes that, on average, can generate a net positive energy. Adapted from Ref. [35]

always on in these solid-state devices, i.e. are independent of time, the engine cycle is reduced to two strokes (see Fig. 6b). The energizing stroke is a partial projective quantum measurement of the WS performed by the spinterface. This instantaneous stroke breaks the quantum entanglement between the two spins and, when averaged over many cycles, can produce a net excess of energy. The second stroke is to (partially) thermalize the WS with the electrodes over a duration time t.

The system's evolution during the thermalizing stroke is thus described by the following master equation [35]:

$$\frac{d\rho}{dt} = -i[H_S, \rho] + \mathcal{T}_L^- \mathcal{D}\left[c_{L\downarrow}^\dagger\right](\rho) + \mathcal{T}_L^+ \mathcal{D}[c_{L\downarrow}](\rho) + \mathcal{T}_R^- \mathcal{D}\left[c_R^\dagger\right](\rho)$$
$$+ \mathcal{T}_R^+ \mathcal{D}[c_R](\rho) + \Lambda^- \mathcal{D}\left[c_{L\downarrow}^\dagger c_{L\uparrow}\right](\rho) + \Lambda^+ \mathcal{D}\left[c_{L\uparrow}^\dagger c_{L\downarrow}\right](\rho).$$

$\mathcal{T}_L^-$ and $\mathcal{T}_R^-$ represent the electron transfer rates between each WS spin and the corresponding left/right electrode, while $\mathcal{T}_L^+$ and $\mathcal{T}_R^+$ correspond to the hole transfer rates. We define the superoperator $\mathcal{D}\left[c_{L\downarrow}^\dagger\right](\rho) = c\rho c^\dagger - \frac{1}{2}\{c^\dagger c, \rho\}$, with $\rho = Tr_E \rho_{SE}$ the WS reduced density matrix compared to the full density matrix ρ_{SE}. The terms

Λ^+ and Λ^- spin represent the boson excitation and relaxation rates when a bosonic bath coupled to the WS is considered. After calculations, the energy change of the nth engine cycle is:

$$\Delta E \equiv \Delta E_n = -TrC\overline{\rho}$$

where $\overline{\rho}$ is the state after the thermalizing stroke of the nth cycle and $C = \gamma c^{\dagger}_{L\downarrow} c_R + \gamma^* c^{\dagger}_R c_{L\downarrow}$ is the tunneling operator that gathers the coherent part of the density matrix. This shows that the partial projective measurement has the effect of releasing the energy of coherence that is contained in the entangled final state of the thermalizing stroke.

So if $\Delta E > 0$, then the QSE should on average output excess energy that could in part be harvested and converted into electrical work. To confirm that this holds in the limit of many cycles, and to examine whether engine operation fluctuations are sufficiently contained to make the result realistic, many engine cycle simulations were performed. As illustrated in Fig. 7a, the temporal/statistical evolution of the QSE first adopts a rapid decay of ΔE, as the purely energetic initial state is increasingly mixed with the thermal state. Then the ensuing stochastic regime is followed by an asymptotic regime that does not reach $\Delta E = 0$. This active device limit behavior depends on the QSE's initial conditions, which suggests that a device initialization sequence may be needed to establish steady-state operation. Exploration of the QSE parameter space suggests the presence of passive, active and chaotic engine states.

Using only quantum measurements as a quantum resource [11] generates non-zero power, even though there is no explicit spin-conserved electron transport path across the junction. This means that Lamblin and Bowen explicitly demonstrate that it is possible to extract energy from quantum fluctuations through the use of autonomous cyclic quantum measurements, while limiting the consideration of important caveats such as the energy of turning on interactions [6].

However, the predicted output power, even at the ~100 THz quantum measurement frequencies calculated [25] by Chowrira et al., remains far below experimental values (~100 nW). This level of output power can be reached if a catalysis from a bosonic bath that directly interacts with the two-spin WS is considered so as to mediate a spin-flip mechanism necessary to promote the tunneling of electrons between the two leads, along with a non-thermal assumption such as a negative temperature bath, a potential or a temperature bias (see Fig. 7b). In that case, thermalization occurs in far fewer cycles, and the QSE is overall more stable against energy fluctuations. This non-thermal bosonic bath is thought to originate from the discrete energy spectrum of phonons, vibrons or magnons on the two-spin WS, which are expected to be present in this solid-state device.

This work raises the final caveat of how to achieve autonomous quantum measurements and the corresponding quantum information erasure. Chowrira et al. [25], and Lamblin and Bowen [35], suggest that the ferromagnetic ground state that underscores the non-thermal spinterface bath act as the measurement apparatus that dissipates information into the electrode behaving as an entropy sink. We discuss in

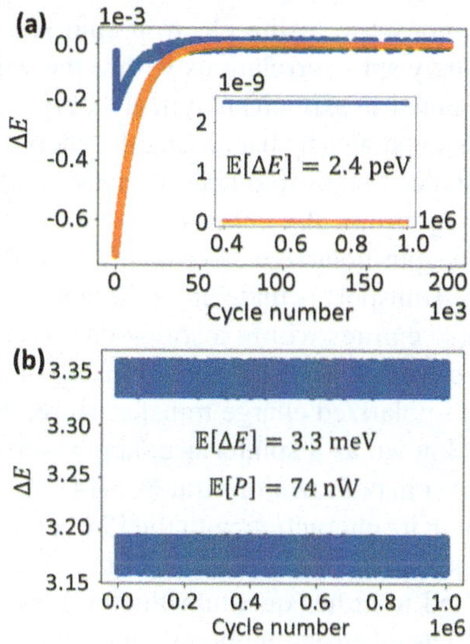

Fig. 7 First formal theory of the Quantum Spintronic Engine: output stability. **a** ΔE *versus* cycle number when only quantum fluctuations are considered, using the initial condition $\rho_0 = |\uparrow\downarrow\rangle\langle\uparrow\downarrow|$. The finite positive ΔE simulated in the final steady state represented in the inset demonstrates that it is possible to harvest energy from quantum fluctuations. **b** ΔE versus cycle number when a negative temperature bosonic bath is included. The data can reproduce the level of output power observed by Chowrira et al. [25] (see Sect. 3.1). Adapted from Ref. [35]

the next section some consequences of this proposal on lifting the final caveat of information erasure, in the form of proposed magnetization dynamical experiments.

4 Conclusions and Perspectives

This Chapter aimed to describe the recently proposed [25, 35] concept of the quantum spintronic engine (QSE), which operates in part using quantum information measurements. We briefly presented the historical background in the field of spintronics that has led to this emergent concept at the intersection with quantum thermodynamics to harvest the energy of quantum and thermal fluctuations using quantum information and other resources as engine fuel. We described the three historical spintronic experiments from the Barnes and Maekawa groups in 2009 [33], from the Moodera group in 2014 [34], and from the Bowen group in 2019 [24] and their similarities/differences with what is presently known about the quantum spintronic engine. These three reports do not propose an explanation in terms of quantum information engines. We then presented the Bowen group's explicit linking of spintronics and quantum information engines. First, Chowrira *et. al* described experiments, *ab-initio* theory and a qualitative concept of the QSE in 2022 [25]. The Chowrira et al. report was the first to explicitly demonstrate, using electron spin resonance, the presence of a quantum resource, namely spin correlations within the working substance. Then, Lamblin and Bowen presented an explicit analytical theory [35], backed by numerical simulations, on a QSE that operates by harvesting the energy of coherence contained within the working substance thanks to quantum measurements.

The QSE operates by spintronically generating an imbalance in the spin channels of transport between fully spin-polarized electrodes across the discrete energy states of paramagnetic centers. Transport is understood here as the steady-state electronic interactions between these entities within a solid-state device. This generates a set of synergetic electronic effects. One is the spin-splitting of the paramagnetic center state due to the fully spin-polarized charge transfer. Here, the ferromagnetic metal/molecule interface, also known as a spinterface, appears to exhibits several advantages over other transport spin polarization strategies [41] for the QSE. Indeed, it can act as a non-thermal bath in its interaction with the PM centers that form the engine's working substance. It enables the measurement of the quantum information on the PM center when subjected to either quantum fluctuations, or thermal fluctuations from a bosonic bath such as electron–phonon coupling. It is thus at the heart of the quantum information engine implementation that is the QSE.

Lamblin and Bowen describe the process of quantum measurement as instantaneous and cyclical. This represents an advantage over other engine strategies [6, 7, 12–14, 16], which ignore the cost of turning on/off interactions. Another notable distinguishing feature is that the coherence energy that is harvested from the measurement inherently arises from the solid-state nature of the electronic interactions that underscore it. Compared to other coherence-driven engine strategies [7, 14], there is no hidden energy accounting to consider apart from the energy cost of the measurement. Another conceptual improvement is to understand the quantum thermodynamical repercussions of the measurement on the ferromagnetic state of the electrode that generates the spinterface's fixed spin referential in the first place. Referring to the work of the Vaccaro group [70], Lamblin and Bowen point out [35] that, aside from energy, another conserved quantity such as spin angular momentum could be used to achieve the required erasure of quantum information, rather than the standard $k_B T \ln 2$ quantum of energy.

This proposal can be tested using well-known spintronic techniques. Indeed, the transfer of spin angular momentum to the ferromagnetic thin film is expected to generation magnetization dynamics that can be detected using ferromagnetic resonance [30]. In this way, the deviation from the ferromagnetic ground state can be quantified, and an effective 'temperature' of spintronic origin established. This would enable an evaluation of the QSE efficiency.

Many aspects of the always-on electronic interactions that characterize the QSE merit further investigation using suitable auxiliary techniques [28]. For example, the spin splitting on the paramagnetic center thanks to the spinterface-induced charge transfer, and the associated energy bandwidth (i.e. stroke frequency) can be evaluated using spin-polarized scanning tunneling microscopy (STM) [26]. An ESR-STM experiment could explicitly examine how spin fluctuations drive changes to the ferromagnetism of the spinterface, and perhaps of the ferromagnetic underlayer. It may even be possible to utilize a STM to assemble a model QSE [28] for which the FM electrode—PM center couplings can be controlled.

Alongside this concept maturation, a renewed effort into materials implementations is underway, but the scale of the R&D investment needs to match the level of ambition that is desired. If the QSE constitutes simply another nifty implementation of quantum thermodynamics engines, then the present intensity of research, which involves only a handful of fundamental researchers, is sufficient. However, as the 2019 Bowen group report emphasized [24], the MgO implementation of the QSE implies that, if control over C, N or Si impurity atoms can be achieved in the MgO barrier of MTJs, then it is in principle possible to utilize the industrial penetration [30] of this spintronic device class to mass produce such energy harvesters of thermal energy, whether naturally present or man-made, with an aerial power density that is competitive compared to other energy strategies such as photovoltaic. It is challenging to control the position of individual atomic impurities in an ultrathin film. Furthermore, technology companies are loathe to change their deposition chain, especially to introduce elements such as C that are essentially a contaminant.

To move quickly on this technology therefore requires a committed investment along the R&D chain. To address our planetary crises, French scientists have called in 2023 for a Manhattan project-like R&D emphasis on high-risk high-gain technology [71]. Should the quantum spintronic engine be one such technology, as Bowen had already proposed back in 2019 [72]?

Acknowledgements We thank C. Elouard and E. Lutz for stimulating discussions. We gratefully acnkowledge PhD funding for M.L. from Ecole Polytechnique. We acknowledge financial support from the ANR (ANR-21-CE50-0039), the Contrat de Plan Etat-Region grants in 2006 and 2008, by "NanoTérahertz", a project co-funded by the ERDF 2014–2020 in Alsace (European Union fund) and by the Region Grand Est through its FRCR call, by the impact project LUEN4S part of the French PIA project "Lorraine Université d'Excellence", reference ANR-15IDEX-04-LUE and by the FEDER-FSE "Lorraine et Massif Vosges 2014–2020", a European Union Program. This 'SpinDrive' work of the Interdisciplinary Thematic Institute QMat, as part of the ITI 2021-2028 program of the University of Strasbourg, CNRS and Inserm, was supported by IdEx Unistra (ANR 10 IDEX 0002), and by SFRI STRAT'US project (ANR 20 SFRI 0012) and EUR QMAT ANR-17-EURE-0024 under the framework of the French Investments for the Future Program.

References

1. Scully, M.O.: Extracting work from a single heat bath via vanishing quantum coherence. Science **299**, 862–864 (2003)
2. Klatzow, J., et al.: Experimental demonstration of quantum effects in the operation of microscopic heat engines. Phys. Rev. Lett. **122**, 110601 (2019)
3. Bresque, L., et al.: Two-qubit engine fueled by entanglement and local measurements. Phys. Rev. Lett. **126**, 120605 (2021)
4. Francica, G., Goold, J., Plastina, F., Paternostro, M.: Daemonic ergotropy: enhanced work extraction from quantum correlations. npj Quantum Inf. **3**, 12 (2017)
5. Elouard, C., Jordan, A.N.: Efficient quantum measurement engines. Phys. Rev. Lett. **120**, 260601 (2018)

6. Jussiau, É., Bresque, L., Auffèves, A., Murch, K.W., Jordan, A.N.: Many-body quantum vacuum fluctuation engines. Phys. Rev. Res. **5**, 033122 (2023)
7. Francica, G., et al.: Quantum coherence and ergotropy. Phys. Rev. Lett. **125**, 180603 (2020)
8. Monsel, J., Fellous-Asiani, M., Huard, B., Auffèves, A.: The energetic cost of work extraction. Phys. Rev. Lett. **124**, 130601 (2020)
9. Buffoni, L., Solfanelli, A., Verrucchi, P., Cuccoli, A., Campisi, M.: Quantum measurement cooling. Phys. Rev. Lett. **122**, 070603 (2019)
10. Rogers, S., Jordan, A.N.: Postselection and quantum energetics. Phys. Rev. A **106**, 052214 (2022)
11. Elouard, C., Herrera-Martí, D., Huard, B., Auffèves, A.: Extracting work from quantum measurement in Maxwell's demon engines. Phys. Rev. Lett. **118**, 260603 (2017)
12. Niedenzu, W., Mukherjee, V., Ghosh, A., Kofman, A.G., Kurizki, G.: Quantum engine efficiency bound beyond the second law of thermodynamics. Nat. Commun. **9**, 165 (2018)
13. Manzano, G., Galve, F., Zambrini, R., Parrondo, J.M.R.: Entropy production and thermodynamic power of the squeezed thermal reservoir. Phys. Rev. E **93**, 052120 (2016)
14. Aimet, S., Kwon, H.: Engineering a heat engine purely driven by quantum coherence (2022). Phys. Rev. A **107**, 012221 (2023). https://journals.aps.org/pra/abstract/10.1103/PhysRevA.107.012221
15. Dago, S., Pereda, J., Barros, N., Ciliberto, S., Bellon, L.: Information and thermodynamics: fast and precise approach to Landauer's bound in an underdamped micromechanical oscillator. Phys. Rev. Lett. **126**, 170601 (2021)
16. Roßnagel, J., Abah, O., Schmidt-Kaler, F., Singer, K., Lutz, E.: Nanoscale heat engine beyond the Carnot limit. Phys. Rev. Lett. **112**, 030602 (2014)
17. Bérut, A., et al.: Experimental verification of Landauer's principle linking information and thermodynamics. Nature **483**, 187–189 (2012)
18. Micadei, K., et al.: Reversing the direction of heat flow using quantum correlations. Nat. Commun. **10**, 2456 (2019)
19. Ji, W., et al.: Spin quantum heat engine quantified by quantum steering. Phys. Rev. Lett. **128**, 090602 (2022)
20. Joachim, C.: The driving power of the quantum superposition principle for molecule-machines. J. Phys.: Condens. Matter **18**, S1935–S1942 (2006)
21. Serri, M., et al.: High-temperature antiferromagnetism in molecular semiconductor thin films and nanostructures. Nat. Commun. **5**, 3079–3087 (2014)
22. Thierschmann, H., et al.: Three-terminal energy harvester with coupled quantum dots. Nat. Nanotechnol. **10**, 854–858 (2015)
23. Koski, J.V., Kutvonen, A., Khaymovich, I.M., Ala-Nissila, T., Pekola, J.P.: On-chip Maxwell's demon as an information-powered refrigerator. Phys. Rev. Lett. **115**, 260602 (2015)
24. Katcko, K., et al.: Spin-driven electrical power generation at room temperature. Commun. Phys. **2**, 116 (2019)
25. Chowrira, B., et al.: Quantum advantage in a molecular spintronic engine that harvests thermal fluctuation energy. Adv. Mater. **34**, 2206688 (2022)
26. Heinrich, A.J., et al.: Quantum-coherent nanoscience. Nat. Nanotechnol. **16**, 1318–1329 (2021)
27. Stevens, J., et al.: Energetics of a single qubit gate. Phys. Rev. Lett. **129**, 110601 (2022)
28. Bowen, M.: Atom-level electronic physicists are needed to develop practical engines with a quantum advantage. npj Quantum Inf. **9**, 25 (2023)
29. Josefsson, M., et al.: A quantum-dot heat engine operating close to the thermodynamic efficiency limits. Nat. Nanotechnol. **13**, 920–924 (2018)
30. Hirohata, A., et al.: Review on spintronics: principles and device applications. J. Magn. Magn. Mater. **509**, 166711 (2020)
31. Tu, S., et al.: Record thermopower found in an IrMn-based spintronic stack. Nat. Commun. **11**, 2023 (2020)
32. Zhang, L., et al.: Ultrahigh detection sensitivity exceeding 105 V/W in spin-torque diode. Appl. Phys. Lett. **113**, 102401 (2018)

33. Hai, P.N., Ohya, S., Tanaka, M., Barnes, S.E., Maekawa, S.: Electromotive force and huge magnetoresistance in magnetic tunnel junctions. Nature **458**, 489–492 (2009)
34. Miao, G.-X., Chang, J., Assaf, B.A., Heiman, D., Moodera, J.S.: Spin regulation in composite spin-filter barrier devices. Nat. Commun. **5**, 3682–3687 (2014)
35. Lamblin, M., Bowen, M.: Quantum measurement spintronic engine powered by quantum coherence enhanced by bosonic catalysis. Phys. Rev. B **109**, 165423 (2024). https://doi.org/10.1103/PhysRevB.109.165423
36. Barthelemy, A., et al.: Magnetoresistance and spin electronics. J. Magn. Magn. Mater. **242**, 68–76 (2002)
37. Ikeda, S., et al.: Tunnel magnetoresistance of 604% at 300K by suppression of Ta diffusion in CoFeB/MgO/CoFeB pseudo-spin-valves annealed at high temperature. Appl. Phys. Lett. **93**, 082508 (2008)
38. Julliere, M.: Tunneling between ferromagnetic films. Phys. Lett. A **54**, 225–226 (1975)
39. Romera, M., et al.: Vowel recognition with four coupled spin-torque nano-oscillators. Nature **563**, 230–234 (2018)
40. Miao, G.-X., Moodera, J.S.: Spin manipulation with magnetic semiconductor barriers. Phys. Chem. Chem. Phys. **17**, 751–761 (2015)
41. Bowen, M.: 16-Spintronics: an application of complex metal oxides. In: Koster, G., Huijben, M. & Rijnders, G. (eds.) Epitaxial Growth of Complex Metal Oxides, 2nd Edn., pp. 469–502. Woodhead Publishing (2022). https://doi.org/10.1016/B978-0-08-102945-9.00009-5
42. Dietl, T.: Zener model description of ferromagnetism in zinc-blende magnetic semiconductors. Science **287**, 1019–1022 (2000)
43. Dietl, T.: A ten-year perspective on dilute magnetic semiconductors and oxides. Nat. Mater. **9**, 965–974 (2010)
44. Fert, A., Jaffrès, H.: Conditions for efficient spin injection from a ferromagnetic metal into a semiconductor. Phys. Rev. B **64** (2001)
45. Djeghloul, F., et al.: High spin polarization at ferromagnetic metal-organic interfaces: a generic property. J. Phys. Chem. Lett. **7**, 2310–2315 (2016)
46. Delprat, S., et al.: Molecular spintronics: the role of spin-dependent hybridization. J. Phys. D Appl. Phys. **51**, 473001 (2018)
47. Misiorny, M., Hell, M., Wegewijs, M.R.: Spintronic magnetic anisotropy. Nat. Phys. **9**, 801–805 (2013)
48. Kent, A.D., Worledge, D.C.: A new spin on magnetic memories. Nat. Nanotech. **10**, 187–191 (2015)
49. Ralph, D.C.: The electromotive force of MnAs nanoparticles. Nature **474**, E6 (2011)
50. Ptaszyński, K.: Autonomous quantum Maxwell's demon based on two exchange-coupled quantum dots. Phys. Rev. E **97**, 012116 (2018)
51. Strasberg, P., Schaller, G., Brandes, T., Esposito, M.: Quantum and information thermodynamics: a unifying framework based on repeated interactions. Phys. Rev. X **7**, 021003 (2017)
52. Cavina, V., Mari, A., Giovannetti, V.: Slow dynamics and thermodynamics of open quantum systems. Phys. Rev. Lett. **119**, 050601 (2017)
53. Chen, X., et al.: Probing superexchange interaction in molecular magnets by spin-flip spectroscopy and microscopy. Phys. Rev. Lett. **101**, 197208 (2008)
54. Campisi, M., Fazio, R.: The power of a critical heat engine. Nat. Commun. **7**, 11895 (2016)
55. Barraud, C., et al.: Unidirectional spin-dependent molecule-ferromagnet hybridized states anisotropy in cobalt phthalocyanine based magnetic tunnel junctions. Phys. Rev. Lett. **114**, 206603 (2015)
56. Barraud, C., et al.: Phthalocyanine based molecular spintronic devices. Dalton Trans. **45**, 16694–16699 (2016)
57. Katcko, K. et al.: Encoding information on the excited state of a molecular spin Chain. Adv. Funct. Mater. 2009467 (2021) https://doi.org/10.1002/adfm.202009467
58. Sánchez, R., Splettstoesser, J., Whitney, R.S.: Nonequilibrium system as a demon. Phys. Rev. Lett. **123**, 216801 (2019)

59. Yurke, B. et al.: Observation of 4.2-K equilibrium-noise squeezing via a Josephson-parametric amplifier. Phys. Rev. Lett. **60**, 764–767 (1988)
60. Chida, K., Nishiguchi, K., Yamahata, G., Tanaka, H., Fujiwara, A.: Thermal-noise suppression in nano-scale Si field-effect transistors by feedback control based on single-electron detection. Appl. Phys. Lett. **107**, 073110 (2015)
61. Hosten, O., Engelsen, N.J., Krishnakumar, R., Kasevich, M.A.: Measurement noise 100 times lower than the quantum-projection limit using entangled atoms. Nature **529**, 505–508 (2016)
62. Wagner, T., et al.: Strong suppression of shot noise in a feedback-controlled single-electron transistor. Nat. Nanotechnol. **12**, 218–222 (2017)
63. Chida, K., Desai, S., Nishiguchi, K., Fujiwara, A.: Power generator driven by Maxwell's demon. Nat. Commun. **8**, 15310 (2017)
64. Lu, Y., et al.: Spin-polarized inelastic tunneling through insulating barriers. Phys. Rev. Lett. **102**, 176801 (2009)
65. Studniarek, M., et al.: Probing a device's active atoms. Adv. Mater. **29**, 1606578 (2017)
66. Fransson, J., Ren, J., Zhu, J.-X.: Electrical and thermal control of magnetic exchange interactions. Phys. Rev. Lett. **113**, 257201 (2014)
67. Jaramillo, J.D.V., Fransson, J.: Charge transport and entropy production rate in magnetically active molecular dimer. J. Phys. Chem. C **121**, 27357–27368 (2017)
68. Ptaszyński, K., Esposito, M.: Thermodynamics of quantum information flows. Phys. Rev. Lett. **122**, 150603 (2019)
69. Bader, K., Winkler, M., van Slageren, J.: Tuning of molecular qubits: very long coherence and spin–lattice relaxation times. Chem. Commun. **52**, 3623–3626 (2016)
70. Wright, J.S.S.T., Gould, T., Carvalho, A.R.R., Bedkihal, S., Vaccaro, J.A.: Quantum heat engine operating between thermal and spin reservoirs. Phys. Rev. A **97**, 052104 (2018)
71. Des scientifiques français appellent « à la mise en œuvre d'un projet Manhattan de la transition écologique. Le Monde 25/09/2023 (2023). https://www.lemonde.fr/idees/article/2023/09/25/nous-appelons-a-la-mise-en-uvre-d-un-projet-manhattan-de-la-transition-ecologique_6190850_3232.html
72. Bowen, M.: Combining spintronics and quantum thermodynamics to harvest energy at room temperature. Phys.org. https://phys.org/news/2019-10-combining-spintronics-quantum-thermodynamics-harvest.html (2019)

Virtual Potential Created by a Feedback Loop: Taming the Feedback Demon to Explore Stochastic Thermodynamics of Underdamped Systems

Salambô Dago, Nicolas Barros, Jorge Pereda, Sergio Ciliberto, and Ludovic Bellon

Abstract Virtual potentials are an elegant, precise and flexible tool to manipulate small systems and explore fundamental questions in stochastic thermodynamics. In particular double-well potentials have applications in information processing, such as the demonstration of Landauer's principle. In this chapter, we detail the implementation of a feedback loop for an underdamped system, in order to build a tunable virtual double-well potential. This feedback behaves as a demon acting on the system depending on the outcome of a continuously running measurement. It can thus modify the energy exchanges with the thermostat and create an out-of-equilibrium state. To create a bi-stable potential, the feedback consists only in switching an external force between two steady values when the measured position crosses a threshold. We show that a small delay of the feedback loop in the switches between the two wells results in a modified velocity distribution. The latter can be interpreted as a cooling of the kinetic temperature of the system. Using a fast digital feedback, we successfully address all experimental issues to create a virtual potential that is statistically indistinguishable from a physical one, with a tunable barrier height and energy step between the two wells.

1 Introduction

Feedback potentials are widely used to trap and manipulate Brownian particles in solution, and explore fundamental questions in non-equilibrium statistical mechanics of small systems [1–4]. Indeed, by controlling an external force acting on a colloidal particle as a function of its measured position, one can create a virtual potential. This is a very powerful tool, more flexible [5] than its physical counterparts consisting of potential forces created by optical or magnetic tweezers [6–10]. Feedback potentials are used in particular to study Landauer's principle, by creating a double well and using the trapped particle as a memory [10, 11]. Within the information processing framework, lowering the dissipation seems a promising path to reduce

S. Dago · N. Barros · J. Pereda · S. Ciliberto · L. Bellon (✉)
Laboratoire de Physique, University of Lyon, ENS de Lyon, CNRS, Lyon, France
e-mail: ludovic.bellon@ens-lyon.fr

X. Bouju and C. Joachim (eds.), *Crossroad of Maxwell Demon*, Advances in Atom and Single Molecule Machines, https://doi.org/10.1007/978-3-031-57904-2_6

energy costs [12–15]. Working with virtual potentials within underdamped dynamics thus appears as a natural endeavor. Moreover, the underdamped regime offers new insights on a wide variety of fundamental questions tackling the connections between feedback and thermodynamics [16–21].

Nevertheless, implementing virtual potentials in the underdamped regime is not an easy task, especially within the stochastic thermodynamics framework that requires to work at the energy scale of $k_B T_0$, where k_B is the Boltzmann constant and T_0 the temperature of the thermal bath. Indeed, resonant systems are very sensitive to perturbations, noises or drifts. Moreover, the feedback update delay can have strong consequences on the coupling between the system and the thermal bath [4, 19].

We describe in this chapter an electrostatic feedback designed to create virtual double-well potentials acting on a micro-cantilever, which serves as an underdamped mechanical oscillator. The feedback acts as a demon, which must be tamed in order to produce a "perfect" virtual potential. Specifically it should not perform nor harvest any work on the system, nor modify the heat exchanges with the thermostat. Although the detailed thermodynamic properties of the demon are not the subject of this study, we show how the non-idealities of the feedback influence the quality of the virtual potential. After this fine tuning, the system offers a high flexibility and precision, with excellent quality in terms of position measurement and force tuning. We are able to create clean, reliable and tunable double-well potentials which outperform those produced by optical and magnetic tweezers (either physical or virtual), and have the added advantage of being analytically tractable. This work therefore presents an unprecedented experimental tool to explore information thermodynamics, and in particular Landauer's principle in the underdamped regime.

In the following, we detail the experimental challenges one faces to remove any bias introduced by the feedback loop. This study incorporates experimental and numerical simulation results, as well as a comprehensive theoretical model. The latter includes the unified and complete description of the switching time of the cantilever in the double-well potential: our expression tends towards Kramer's escape time in the high-energy barrier limit, but it also provides an adjusted model for barriers lower than the thermal energy, where Kramer's formula is no longer valid.

The chapter is organised as follows: we first present the experimental system and the principle of the feedback loop to create a virtual potential indistinguishable from an equivalent physical one (Sect. 2). We then explore the non-idealities of an experimental implementation (Sect. 3). In particular, we analyse how measurement noise or delays in the switches between the wells result in a bias of the energy exchanges with the thermal bath, effectively warming or cooling the oscillator Brownian noise. From this comprehensive analysis, we describe the digital filtering process with anticipation we need to implement to mitigate imperfections. Lastly, Sect. 4 describes how we can use the feedback loop as a demon to induce a controlled heat flow between the thermal bath and the system, driving it into a non-equilibrium steady state. We finally conclude on the potential of this approach to explore the demon behavior from the information prism.

2 Virtual Double Well Potential: Principle

In this section we describe first the experimental set-up whose main component is a cantilever behaving like an harmonic oscillator. We then present how the feedback force acts on the system and how it is calibrated in order to perform quantitative measurements and to compare the experimental results to the theoretical ones. We finally describe how double well potentials can be created, and the amount of energy exchanged due to the feedback when crossing the barrier from one well to the other.

2.1 From a Simple Oscillator to a Bistable System

As sketched in Fig. 1, the underdamped oscillator is a conductive cantilever which is weakly damped by the surrounding air at room temperature T_0. Its deflection x is measured with very high signal-to-noise ratio by a differential interferometer [22]. The Power Spectral Density (PSD) of the thermal fluctuations of x is plotted in Fig. 2: the fundamental mode dominates by 3 orders of magnitude the higher-order deflection modes of the cantilever. The second deflection mode at 8 kHz is conveniently removed from the measured signal by focusing the sensing laser beam on its node, at around 0.78% of the cantilever length. This adjustment helps in having a physical system very close to an ideal Simple Harmonic Oscillator (SHO), characterized by its resonance frequency $f_0 = \omega_0/2\pi$, mass m, stiffness $k = m\omega_0^2$ and quality factor Q. The fit of this PSD with the theoretical thermal noise spectrum of a SHO leads to

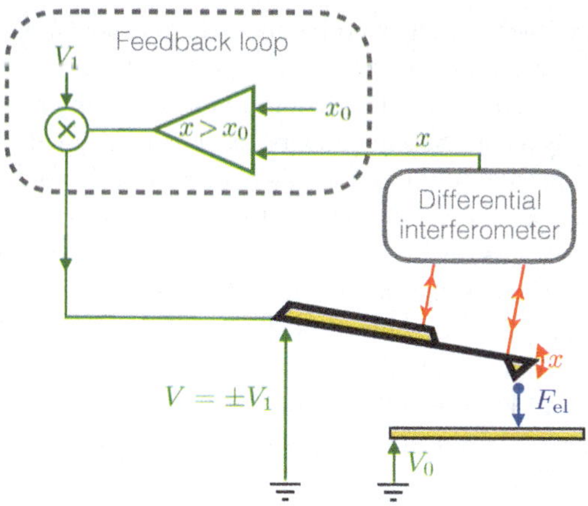

Fig. 1 Experimental system. The conductive cantilever is sketched in yellow. Its deflection x is measured with a differential interferometer [22], by two laser beams focused on the cantilever and on its base. The cantilever at voltage $V = \pm V_1$ is facing an electrode at V_0. The voltage difference $V - V_0$ creates an attractive electrostatic force $F \propto (V - V_0)^2$. The double-well potential is created by the feedback loop, consisting in a comparator and a multiplier (dashed box)

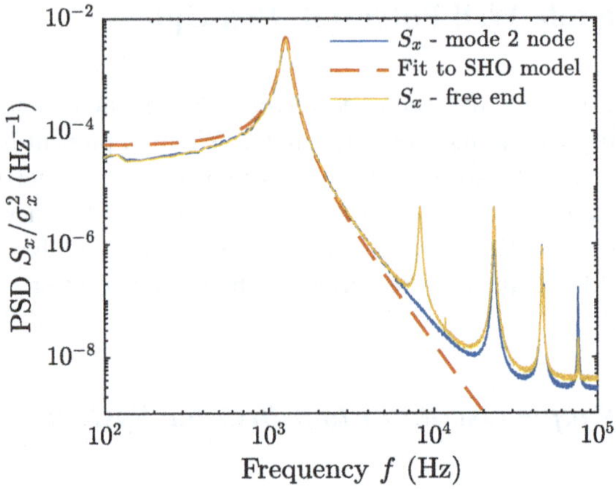

Fig. 2 **PSD S_x of the thermal noise driven deflection in a single well.** S_x is measured with no feedback ($V_1 = 0$, solid lines), fitted by the theoretical spectrum of a SHO (dashed line). The second deflection mode, visible at 8 kHz when the laser beam is focused at the free end of the cantilever (yellow), is successfully hidden by focusing the laser beam on the node of this mode (blue). Up to 10 kHz, the cantilever behaves like a SHO at $f_0 = 1270$ Hz, with a quality factor $Q = 10$

$f_0 = 1270$ Hz, $k \sim 4 \times 10^{-3}$ N/m and $Q \sim 10$ in air. The slight difference between the measurement and the model is due to frequency dependence of the viscous damping of the cantilever in air [23, 24]. Higher quality factors can be achieved in vacuum. We used as the length scale the variance at equilibrium $\sigma_x^2 = k_B T_0/k \sim 1\,\text{nm}^2$.

As such, the cantilever is an underdamped oscillator moving in an harmonic potential of stiffness fixed by the spring constant of the mechanical beam. In order to use the cantilever in stochastic thermodynamics experiments, we need to be able to tune this potential. For example to use it as a one-bit memory and probe Landauer's bound [12, 13, 25], we need to confine its motion in an energy potential consisting of two wells separated by a barrier, whose shape can be tuned at will. Physical potentials with non-linearities at the $k_B T_0$ scale are very difficult to create for such an object. We therefore turn towards a virtual potential U created by a feedback loop.

In this chapter, we focus on the creation of a double well potential, and implement a feedback which compares the cantilever deflection x to an adjustable threshold x_0. After having multiplied the output of the comparator by an adjustable voltage V_1, the result is a feedback signal $V = S(x - x_0)V_1$, with $S(\cdot)$ the sign function: $V = +V_1$ if $x > x_0$ and $V = -V_1$ if $x < x_0$. The voltage V is applied to the cantilever which is at a distance d from an electrode kept at a voltage V_0. The cantilever-electrode voltage difference $V_0 \pm V_1$ creates an electrostatic attractive force $F = \frac{1}{2}\partial_d C(d)(V_0 \pm V_1)^2$ [26], where $C(d)$ is the cantilever-electrode capacitance. Since $d \gg \sigma_x$, $\partial_d C(d)$ can be assumed constant. We apply $V_0 \sim 100$ V and $V_1 \ll V_0$ so that, to a good approximation, $F \propto \pm V_1$ up to a static term. This feedback loop results in the application of an external force whose sign depends on whether the cantilever is above or below the threshold x_0. If the feedback loop is fast enough, the switching transient is negligible. As a consequence, the oscillator evolves in a virtual static double-well potential, whose features are controlled by the two parameters x_0 and V_1. Specifically, the barrier position is set by x_0 and its height is controlled indirectly by V_1, which sets the wells centers $\pm x_1 = \pm V_1 \partial_d C(d) V_0/k$. The potential energy constructed by this feedback is computed by integrating the total force (spring + electrostatic) as a function of position:

$$U(x, x_0, x_1) = \frac{1}{2}k\big(x - S(x - x_0)x_1\big)^2 + kx_0x_1\big(S(x - x_0) + S(x_0)\big). \quad (1)$$

2.2 The Virtual Double Well Potentials

The potential energy in Eq. (1) can be experimentally measured from the Probability Distribution Function PDF(x) and the Boltzmann equilibrium distribution: PDF$(x) \propto e^{-U(x)/k_B T_0}$. Figure 3 presents examples of experimental (a) symmetric and (b) asymmetric double-well potentials generated by the feedback loop we describe in this chapter. The dashed lines are the best fits with Eq. (1), demonstrating that the feedback-generated potential behaves as a static one, in terms of the position PDF.

The second degree of freedom of the underdamped system is the velocity $v = \dot{x}$, and it is also expected to satisfy the Boltzmann equilibrium distribution. The equilibrium PDF of the velocity in the double-well should be the same as the one in a single harmonic well: a Gaussian of variance $\sigma_v^2 = k_B T_0/m$ PDF$(v) \propto e^{-mv^2/(2k_B T_0)}$. As shown in Fig. 3c, our implementation of the feedback also validates this point.

In the following, we present the strategy we implemented to create such a virtual potential. Using such a fine tuned feedback loop, or good behaving demon, the potential is identical to a physical one, without any noticeable effect on the position and velocity equilibrium distributions.

2.3 Energy Exchanges When Switching Wells

The main principle behind the virtual potential construction is to use a demon that switches the harmonic well center exactly when x crosses the threshold x_0. Let us analyse here the upward switch from $-x_1$ to $+x_1$, the extension to the other case is straightforward. If the demon acts right on time with a negligible switching time, then the oscillator evolves in the desired potential defined by Eq. 1, and no energy is exchanged during the switch: the demon is invisible with respect to the defined potential.

However, the demon acts according to the available information, that is the measured position x_m, which can be noisy, or somewhat delayed with respect to the actual position of the oscillator x. It means that in practice, the demon switched

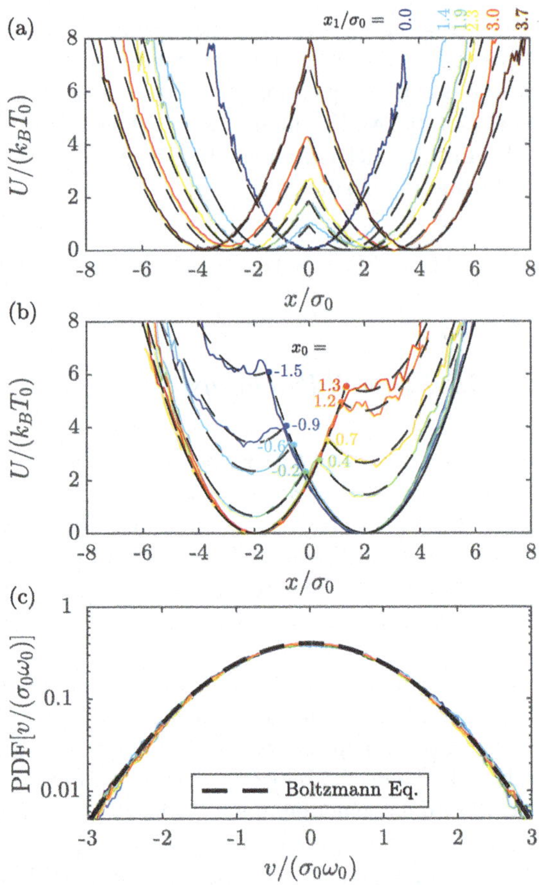

Fig. 3 **(a)Virtual symmetric doubled well potentials** inferred from the position PDF and the Boltzmann equilibrium distribution, for half-distances $x_1 = 0$ to $3.7\sigma_x$ and a barrier position at $x_0 = 0$. Dashed line are best fits to Eq. 1. **(b) Virtual asymmetric doubled well potentials** for a half-distance $x_1 = 2\sigma_x$ and barrier positions from $x_0 = -1.5$ to $1.3\sigma_x$. Dashed line are best fits to Eq. 1. **(c) Velocity PDFs** corresponding to the data of (a): they are indistinguishable from the one corresponding to the equilibrium in a single well. All the data presented in this figure is acquired in vacuum with a quality factor $Q = 80$

Fig. 4 **Energy associated with an error on the threshold**: if instead of switching wells when x crosses x_0, the demon acts at $x_0 + h$ with $h > 0$, the right well is a bit higher than expected, by $\Delta U = 2kx_1h$

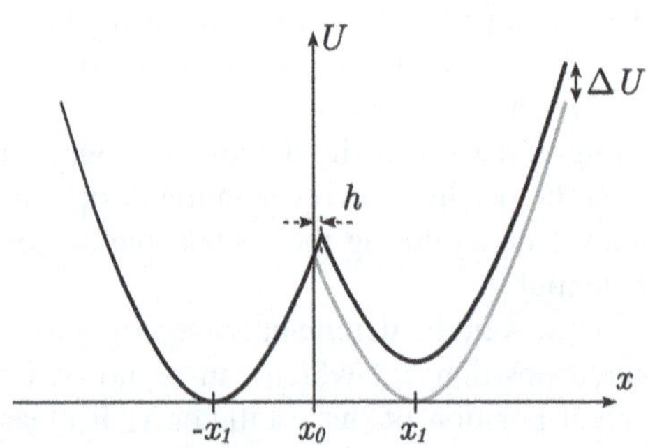

when the oscillator crossed $x = x_0 + h$, with h a small quantity describing the error with respect to the desired threshold. Right after switching, the oscillator is thus evolving in the potential $U(x, x_0 + h, x_1)$, which is shifted by $\Delta U(h) = 2kx_1h$ with respect to the goal potential (illustration Fig. 4). This ΔU can be understood as the energy taken or given by the demon (depending on the sign of h) to the oscillator, supposed to be evolving in the potential $U(x, x_0, x_1)$. To further illustrate this point, let us consider the kinetic energy of the mechanical system, assuming a high quality factor, thus low energy exchanges with the thermostat on short time scale. If $h > 0$, the oscillator had to climb the initial potential a bit higher than anticipated before switching to the next well, thus loosing the kinetic energy equivalent to $U(x_0 + h, x_0 + h, x_1) - U(x_0 + h, x_0, x_1) = 2kx_1h$: the oscillator is somewhat colder than it should have, this energy was taken by the demon from the system. If $h < 0$, the situation is opposite and the demon gives energy to the oscillator which get somewhat hotter.

When analyzing the downward switch from $+x_1$ to $-x_1$, the conclusion are equivalent, up to a sign reversal for h: $\Delta U(h) = -2kx_1h$. $h > 0$ is anticipating the switch and warms the system, while $h < 0$ is delaying the switch and pumps energy from the system. Upon multiple crossing of the barrier, all these energy exchanges add up and can create a net energy flux to or from the system to the demon. If h is a deterministic or random variable independent of the direction, then in average the net flux is zero and the demon is transparent: the effective mean potential is indistinguishable from a physical one.

3 Digital Feedback Loop Implementation

In this section we describe the task of performing the switch between wells at the right time. Noise creates early and repeated switches which should be avoided (wrong virtual potential, warming), so we need to filter the high frequency noise. This process induces a delay in the response, resulting in an effective alternating asymmetric double wells which in turns cools the system. So we need to anticipate the signal to compensate for this delay and finally reach our goal. After describing this strategy, we present how to implement it in a digital feedback loop, going into the details of the technical steps of the filter, and of the resulting transfer function.

3.1 Taming the Demon

The task of the demon inside the feedback loop is to switch the harmonic well center to the right position $\pm x_1$ whenever the position signal x crosses the threshold x_0. As simple as this operation may look like, the real world implementation is not straightforward. Indeed, as illustrated by the power spectral density of the signal plotted in Fig. 2, on top of the signal of interest x (the first mode of the oscillator),

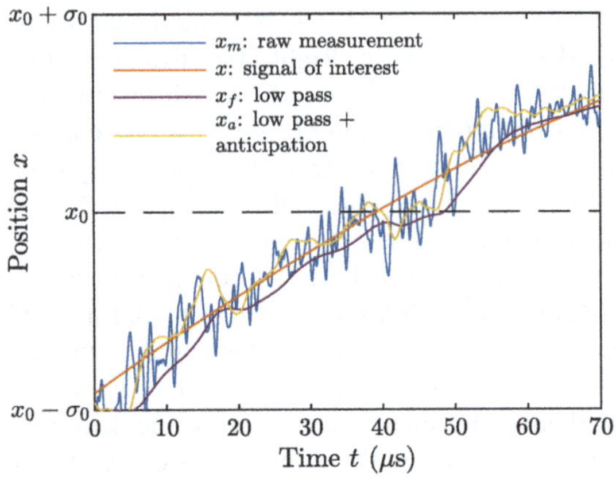

Fig. 5 Triggering right on the threshold: the measured signal x_m is hampered by high frequency noise with respect to the signal of interest x. This noise creates detrimental multiple and early triggering. Standard low pass filtering delays the signals and thus triggers late. We implement a low pass filter with an anticipation based on the speed to detect the crossing of x_0 on time in average

high frequency noise is present in the measured signal x_m. This noise originates from higher order resonance modes (mode 3 and above, thanks to the trick of canceling the second mode with the proper choice of the probe laser) of the cantilever, as well as the floor noise of the interferometer due to the shot noise of the photodiodes.

As illustrated in Fig. 5 on synthetic signals, this high frequency noise has an important consequence on the comparator: repeated high frequency commutation between the two wells. This effect is counterproductive in our case: during fast and repeated switches between $\pm V_1$, the average voltage experienced by the cantilever is typically 0, therefore the mode 1 of the oscillator responding only to low frequencies can be stabilized in a well centered around the origin. This effect is generally canceled in standard comparators by introducing an hysteresis $h > 0$ larger than the noise amplitude: once triggered ($x_m > x_0 + h$ for instance), x_m is sufficiently far from the new threshold $x_0 - h$ to trigger the comparator again. This approach is however forbidden in our case: it would result in a systematic pumping of energy $\Delta U = 2kx_1h > 0$ at each barrier crossing in both directions, thus effectively cooling the system. Such a demon would result in a constant heat flow from the oscillator, creating a non-equilibrium steady state and modifying the distribution of velocities.

As an alternative, we implement a temporal lock-up to freeze the comparator state after a switch, for 1/4 of the oscillator's natural period $1/f_0$. By the time the comparator is active again, the cantilever has evolved in the new well on average long enough to reach the bottom of the well, and is therefore far enough from the threshold that an undue noise-induced switch is improbable. One drawback is that short excursions in the other well are forbidden as well. However these events, indeed present in a real double-well potential, are unlikely enough that removing them has no noticeable effect of the statistical properties of the virtual potential.

This temporal lock-up strategy only partially solves the problem of high frequency noise. Indeed, this noise results in an early trigger, since the measured signal x_m always presents a fluctuation crossing the threshold before x does. It corresponds to an effective negative hysteresis $h < 0$, which results in a systematic injection of energy $\Delta U = 2kx_1h < 0$ when crossing the barrier in both directions, thus effectively heating the system. For example, the thermal noise amplitude of mode 3 alone is $4 \times 10^{-2}\sigma_x$, corresponding to $\Delta U \sim -0.08k_BT_0$ at each crossing for a barrier height $\mathcal{B} = \frac{1}{2}k_BT_0$. Again, such a demon would result in a non-equilibrium steady state with a constant heat flow to the oscillator.

We therefore need to filter these high frequency fluctuations, using a low pass filter cutoff frequency f_c below mode 3 resonance $f_3 = 23\,\text{kHz}$, choosing for example $f_c = 10\,\text{kHz}$. However, this filtering induces a phase delay ϕ via its transfer function. For the third order Butterworth filter that we use, this phase is $\phi_{\text{BW}_3} \sim 2f/f_c \sim 0.1$ in the frequency range of interest, close to the resonance of mode 1. The filtered signal x_f is therefore delayed by $\tau_d \sim 1/(\pi f_c) = 32\,\mu\text{s}$ with respect to x. Since the typical speed when crossing the threshold is $\sigma_v = \sigma_x\omega_0$, the delay corresponds to an average positive hysteresis $h = \sigma_v\tau_d = 2\sigma_x f_0/f_c \sim 0.2\sigma_x$, thus to $\Delta U \sim 0.4k_BT_0$ at each crossing for $x_1 = \sigma_x$. One more time, such a demon would result in a constant heat flow from the oscillator, creating a non-equilibrium steady state.

However, since this delay τ_d due to the filter is known, one can measure the actual speed of the oscillator and anticipate when the value of x will cross the threshold, and thus trigger in average right on time. This strategy is illustrated in Fig. 5: the speed, computed on the filtered signal $x_f(t)$, is use to construct $x_a(t) = x_f(t) + \tau_a\dot{x}_f(t) \approx x_f(t + \tau_a)$. Since $x_f(t) \approx x(t - \tau_d)$, $x_a(t) \approx x(t - \tau_d + \tau_a)$ can be a good forecast of the signal of interest choosing $\tau_a = \tau_d$. This is the final strategy that we implement in our feedback loop, and we adjust the value of τ_a of the experimental setup to compensate both for filtering and additional delays induced by the total feedback loop. Such taming of our demon results in the data displayed in Fig. 3, where the virtual potential is indiscernible from a physical one, both in position and speed.

3.2 *Practical Implementation*

To allow for fast and versatile strategies, our feedback loop is implemented using a FPGA (field-programmable gate array) board associated with fast analog to digital (ADC) and digital to analog (DAC) circuits. A Labview-based FPGA data acquisition system from NI (FPGA 7975R + Adapter module NI 5783) allows operations clocked at 10 ns (100 MHz acquisition frequency, 200 MHz onboard clock), with the ADC and DAC operations (16 bits) performed in approximately 500 ns. The onboard operations are the following:

1. ADC: sampling of the 4 photodiodes signal of the interferometer. Those signals of amplitude around 0.5 V are collected at the output of analog current to voltage amplifiers with a gain of 10^5 V/A and a bandwidth of 1 MHz [22].

2. Computing of the deflection x_m of the cantilever. This step implies an arctan function, and multiplication by pre-calibrated coefficients, and last 270 ns (27 clock ticks).
3. Digital filtering. Using a single digital IIR filter (infinite impulse response) combining a third order Butterworth filter and an anticipation by τ_a of the position, we compute x_a. This steps is 100 ns long (10 clock ticks).
4. Comparison to the threshold. x_0 is extracted from a lookup table (to allow time dependent protocols), and compared to x_a. According to the result, a voltage V_1 or $-V_1$ is selected from a second lookup table.
5. DAC: the voltage V_1 is sent to the device output.
6. Export data: in parallel to the feedback operations, x_m, x_0 and the output voltage $V = \pm V_1$ are low pass filtered (antialiasing cutoff frequency 670 kHz of an independent IIR filter), down-sampled at 2 MHz, and exported to the host computer for post-processing.

The total deterministic delay to be anticipated by these operations is $\tau_d \sim 1\,\mu$s. Some details of each steps are described in the next paragraphs.

3.2.1 Before the Experiment: Calibration

We first perform a calibration step implementing in the FPGA the calibration coefficients to convert the interferometer four photodiodes voltage signals to an actual position x in nm [22]. We then perform a ramp in the voltage V applied to the tip and read the cantilever average position $\langle x_m \rangle$, to convert V_1 into x_1 and vice-versa. The origin of the x axis is periodically (every protocol) set to $(\langle x \rangle = 0, V = 0)$ in order to remove the drift in position during long experiments.

3.2.2 Filtering

The raw signal x_m should be low pass filtered ($x_f = \mathrm{BW}_3 \otimes x_m$, with BW_3 a third order Butterworth filter of low pass cutoff frequency $f_c = 10$ kHz) and anticipated ($x_a = x_f + \tau_a \dot{x}_f$) before being compared to the threshold x_0. In the frequency space, those operations correspond to the following response function:

$$x_a(\omega) = (1 + i\omega\tau_a)x_f(\omega) = \frac{1 + i\omega\tau_a}{(1 + i\omega\tau_c)(1 + i\omega\tau_c - \omega^2\tau_c^2)} x_m(\omega) \quad (2)$$

$$= \frac{\sum_{k=0}^{3} d_k (i\omega\tau_c)^k}{\sum_{k=0}^{3} c_k (i\omega\tau_c)^k} x_m(\omega) \quad (3)$$

where $\tau_c = 1/(2\pi f_c)$, $C = [c_0, c_1, c_2, c_3] = [1, 2, 2, 1]$ and $D = [d_0, d_1, d_2, d_3] = [1, \tau_a/\tau_c, 0, 0]$. In the discrete time domain where the FPGA operates, this filter should be translated in operations on the sampled values $x_m(t_n)$ at each time sampling $t_n = n\delta t$, with $\delta t = 10$ ns. We implement this filter using an IIR scheme: the filter output at time t_n is computed from the previous value of the output and of the input, as:

$$x_a(t_n) = \frac{1}{a_0}\left(\sum_{k=0}^{3} b_k x_m(t_{n-k}) - \sum_{k=1}^{3} a_k x_a(t_{n-k})\right) \tag{4}$$

where the coefficients $A = [a_k]_{k=0 \text{ to } 3}$ and $B = [b_k]_{k=0 \text{ to } 3}$ have to be inferred from the desired frequency response of Eq. 3. Noting that $i\omega x(\omega)$ in the frequency domain corresponds to the discrete derivative $[x(t_n) - x(t_{n-1})]/\delta t$ in the discrete time domain, we construct the translation from one domain to the other as $A = MC$ and $B = MD$, where the matrix M is defined by

$$M = \begin{bmatrix} 1 & \alpha & \alpha^2 & \alpha^3 \\ 0 & -\alpha & -2\alpha^2 & -3\alpha^3 \\ 0 & 0 & \alpha^2 & 3\alpha^3 \\ 0 & 0 & 0 & -\alpha^3 \end{bmatrix} \tag{5}$$

with $\alpha = \tau_c/\delta t$. From the desired frequency response (coefficients C and D), one can compute the IIR coefficients A and B to implement in the FPGA programming.

When following this strategy, we run into two sources of instabilities for the IIR filters. The first one is due to the the very low cutoff frequency $f_c = 10\,\text{kHz}$ with respect to the sampling frequency $f_s = 100\,\text{MHz}$: these 4 orders of magnitude makes the operation of Eq. 4 very sensitive to rounding errors, has we compute $x_a(t_n)$ from the difference of very large numbers compared to the result. We therefore need to perform 64 bits operations during the filtering process. The increased computational cost translates into a second source of instability of the IIR filter: each multiplication in Eq. 4 requires 9 FPGA ticks. As such, at time t_n the information available is t_{n-9} and not t_{n-1}. We therefore need to rewrite Eq. 4 as:

$$x_a(t_n) = \frac{1}{a'_0}\left(\sum_{k=0}^{3} b'_k \bar{x_m}(t_{n-9k}) - \sum_{k=1}^{3} a'_k x_a(t_{n-9k})\right) \tag{6}$$

where the coefficients $A' = [a'_k]_{k=0 \text{ to } 3}$ and $B' = [b'_k]_{k=0 \text{ to } 3}$ are now computed with $\alpha' = \tau_c/(9\delta t)$. To avoid running 9 independent IIR filters in parallel, we actually use as an input the average signal $\bar{x_m}$ obtained with a sliding rectangular window on 9 samples:

$$\bar{x_m}(t_n) = \frac{1}{9}\sum_{k=1}^{9} x_m(t_{n-k}). \tag{7}$$

Formally, this is a low pass FIR (finite impulse response) filter applied to x_m before the IIR filter, with a cutoff frequency of $f_s/9 = 11\,\text{MHz}$.

Implementing all those bricks results in a stable and efficient digital filter, designed to remove high frequency fluctuations and correct for the filter dephasing and the $\sim 1\,\mu\text{s}$ feedback delay. Formally, we would like the overall phase of FPGA feedback

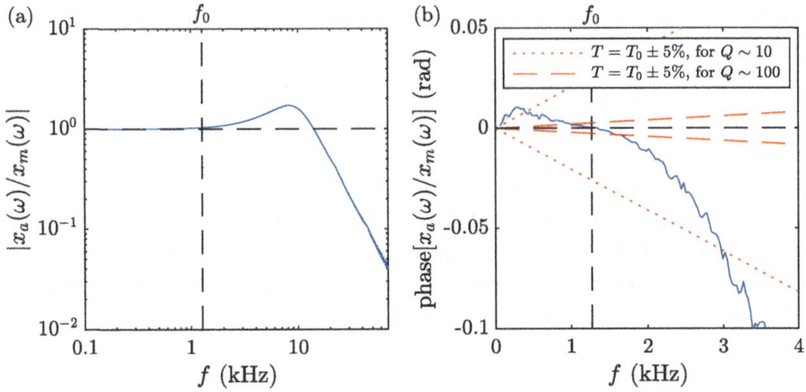

Fig. 6 **(a) Gain of the FPGA position signal conditioning**. As expected the gain of the feedback operations on the position signal (ADC, conversion of the photodiode outputs into x signal, IIR filter and DAC) is 1 (horizontal dashed black) within the frequency range of interest (f_0 in vertical dashed black), and is then filtered above $f_c = 10\,\text{kHz}$ to remove the 3rd deflection mode contribution. We notice that there is a slight amplification above f_0 due to the resonance of the IIR filter. **(b) Phase of the FPGA position signal conditioning**: the IIR filter is designed to have a 0 phase for the overall processing of the position signal at f_0 (vertical dashed black), to avoid switching advance or delay, and hence meet the $T = T_0 \pm 5\%$ requirement on the kinetic temperature. The authorized range (computed with Eq. 34) corresponds to the interval between the red dashed lines for $Q \sim 100$, and between the dotted lines for $Q \sim 10$

to vanish around the oscillator's resonance. In practice, the anticipation coefficient τ_a is fine tuned experimentally to enforce an unbiased velocity PDF in the resulting double-well potential. The Bode diagram of the overall FPGA operations on the position signal (ADC, conversion of the photodiode outputs into x_m signal, FIR and IIR filter, DAC) is displayed in Fig. 6. It consists in the gain and the phase of the transfer function between the real time position directly inferred from the photodiode outputs (no real time conversion, nor filtering, but only post treatment operations), and the position signal acquired, reconstructed, filtered and output by the FPGA device, x_a. It is worth noticing on Fig. 6a that this real time zero-phase -at f_0- filtering is performed at the expense of a very small resonance in the filter gain, which has negligible consequences on our experiments.

3.3 Recording and Post-treatment Analysis

To compute the thermodynamics quantities we use the measured position x_{acq} properly filtered above 10 kHz in post-treatment analysis using a zero-phase filtering (processing the data in both the forward and reverse direction). This post-acquisition filter is better than the real time FPGA one (because it enforces 0 phase for all frequencies and not only f_0), that is why we do not use directly x_f for the analysis. Let us note that such zero phase filtering is not causal and thus cannot be implemented on the FPGA for real time operations. Thanks to the calibration of the data, no further post-treatment is required.

4 Creating Non-equilibrium Steady States with Unleashed Demons

In this section we explore the properties of the steady state when the total feedback delay is not tuned to zero: each time we cross the barrier, we lose or gain energy. Therefore we need to compute the barrier crossing rate, and from this we can deduce the effective kinetic temperature of the system, that we compare to the experiment.

4.1 Beyond Kramer's Escape Rate: Switching Between Wells Separated by a Low Barrier

Let us study a generic one degree of freedom mechanical system of mass m and position x evolving in a potential energy $\bar{U}(x)$. Its kinetic energy is $K = p^2/(2m)$, with $p = mdx/dt$ the momentum, so that its Hamiltonian is:

$$H(x, p) = \frac{1}{2m}p^2 + \bar{U}(x). \tag{8}$$

We suppose it is statistically at temperature T: the initial conditions in position and momentum are drawn from a Boltzmann distribution

$$P(x, p) = \frac{1}{Z}e^{-\beta H(x,p)}, \tag{9}$$

$$Z = \iint dx\, dp\, e^{-\beta H(x,p)}, \tag{10}$$

with $\beta = 1/(k_B T)$.

We suppose that the potential presents a double well, as sketched in Fig. 7. The barrier between the two wells is placed in $x = 0$ and its energy is $\bar{U}(0) = B$, with the origin of $\bar{U}$ at its minimum. In the limit of weak damping, the total energy $E = H(x, y)$ of the system is conserved in time. Therefore, its Hamiltonian dynamics can be solved for a given total energy E by integrating the following equation:

$$\frac{dx}{dt} = \pm\sqrt{\frac{2}{m}[E - \bar{U}(x)]}. \tag{11}$$

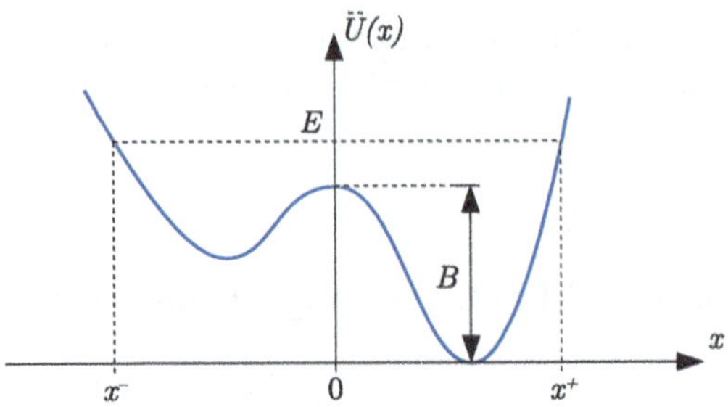

Fig. 7 We study the escape rate from one well ($x > 0$) to the other ($x < 0$) on a generic double well potential $\bar{U}(x)$. The barrier height is B, and is positioned at $x = 0$. For a total energy $E > B$ of an Hamiltonian system evolving in this potential, the motion is periodic between x^- and x^+. For $E < B$, the system is stuck in the well of its initial condition

The motion is periodic: the time $\bar{\mathcal{T}}(E)$ to explore the full phase space available for a given E, depends on the value of E with respect to the barrier height $\mathcal{B}$:

- If $E > \mathcal{B}$ the system explores both wells every period, so there are 2 barrier crossing (one in each direction) every period, with the period being:

$$\bar{\mathcal{T}}(E > \mathcal{B}) = 2\int_{x^-}^{x^+} \frac{dx}{\sqrt{2[E - \bar{U}(x)]/m}}, \tag{12}$$

  where x^- and x^+ are the minimum and maximum values of x allowed by the criterion $\bar{U} = E - K \leq E$.
- If $E < \mathcal{B}$, then the motion is confined to a single well, there are no switches, there are actually 2 periods $\bar{\mathcal{T}}^-$ and $\bar{\mathcal{T}}^+$ to consider depending on which well is explored given the initial conditions.

We define in the following the time spent in each well on a period as:

$$\bar{\mathcal{T}}^-(E) = 2\int_{x^-}^{0} \frac{dx}{\sqrt{2[E - \bar{U}(x)]/m}}, \tag{13}$$

$$\bar{\mathcal{T}}^+(E) = 2\int_{0}^{x^+} \frac{dx}{\sqrt{2[E - \bar{U}(x)]/m}}, \tag{14}$$

where again the bound 0 of the integrals should be adjusted when $E < \mathcal{B}$. When $E > \mathcal{B}$, we trivially verify $\bar{\mathcal{T}}(E) = \bar{\mathcal{T}}^-(E) + \bar{\mathcal{T}}^+(E)$, and we extend this relation to $E < \mathcal{B}$ to define the time to explore the full space available for a given energy.

As an illustration, we can use the symmetric bi-quadratic potential $U = \frac{1}{2}k(|x| - x_1)^2$, corresponding to $x_0 = 0$ and $\mathcal{B} = \frac{1}{2}kx_1^2$. We compute

$$\text{for } E < \mathcal{B} : \tfrac{1}{2}\mathcal{T}(E) = \mathcal{T}^{+}(E) = \mathcal{T}^{-}(E) = \mathcal{T}_0, \tag{15a}$$

$$\text{for } E > \mathcal{B} : \tfrac{1}{2}\mathcal{T}(E) = \mathcal{T}^{+}(E) = \mathcal{T}^{-}(E) = \mathcal{T}_0 \left[\frac{1}{2} + \frac{1}{\pi} \sin^{-1}\left(\sqrt{\frac{B}{E}} \right) \right] \tag{15b}$$

$\mathcal{T}(E)$ is twice the natural period $\mathcal{T}_0 = 2\pi/\omega_0$ of a single oscillator when $E \leq \mathcal{B}$ or $E \gtrsim \mathcal{B}$, and tends to $\mathcal{T}_0$ for $E \gg \mathcal{B}$.

Since time is a natural variable for this problem, we operate a change of variables from (x, p) to (t, E): for any given position and momentum, we can define a unique time and energy, and vice-versa. We consider here that as the system is periodic, the time range is restricted from $-\bar{\mathcal{T}}^{-}(E)$ to $+\bar{\mathcal{T}}^{+}(E)$ for a given total energy E. This is a canonical transformation for the Hamiltonian since the Jacobian determinant is 1:

$$dx\, dp = \left| \frac{\partial x}{\partial E}\frac{\partial p}{\partial t} - \frac{\partial x}{\partial t}\frac{\partial p}{\partial E} \right| dt\, dE = dt\, dE, \tag{16}$$

The change of variable is therefore straightforward for the probability:

$$P(t, E) = \frac{1}{Z} e^{-\beta E}, \tag{17}$$

where $P(t, E)$ is defined only for $-\bar{\mathcal{T}}^{-}(E) < t < \bar{\mathcal{T}}^{+}(E)$. The probability of having an energy E is deduced by integrating on time, leading to:

$$P(E) = \frac{1}{Z} e^{-\beta E} \bar{\mathcal{T}}(E), \tag{18}$$

$$\text{with } Z = \int_0^{\infty} dE\, e^{-\beta E} \bar{\mathcal{T}}(E). \tag{19}$$

This computation can be done while limiting the explored space to the left or right well, leading to the same expression where we only substitute $\bar{\mathcal{T}}(E)$ by $\bar{\mathcal{T}}^{\pm}(E)$: $P^{\pm}(E) = e^{-\beta E}\bar{\mathcal{T}}^{\pm}(E)/Z^{\pm}$, with $Z^{\pm} = \int dE\, e^{-\beta E}\bar{\mathcal{T}}^{\pm}(E)$.

We can now compute the escape rate from one well to the other, lets' say Γ^{+} from $x > 0$ to $x < 0$. For a given energy E, the escape rate is 0 if $E < \mathcal{B}$, and $1/\bar{\mathcal{T}}^{+}(E)$ otherwise (the system spend one period in the well before switching to the other, so one crossing every period). The mean transition rate is therefore

$$\Gamma^{+} = \int_{\mathcal{B}}^{\infty} dE \frac{1}{\bar{\mathcal{T}}^{+}(E)} P^{+}(t, E) = \int_{\mathcal{B}}^{\infty} dE \frac{1}{\bar{\mathcal{T}}^{+}(E)} \frac{1}{Z^{+}} e^{-\beta E} \bar{\mathcal{T}}^{+}(E) \tag{20}$$

$$= \frac{\int_{\mathcal{B}}^{\infty} dE\, e^{-\beta E}}{\int_0^{\infty} dE\, e^{-\beta E} \bar{\mathcal{T}}^{+}(E)} = \frac{e^{-\beta \mathcal{B}}}{\int_0^{\infty} dE \beta e^{-\beta E} \bar{\mathcal{T}}^{+}(E)}. \tag{21}$$

This expression is generic and applies to any double well potential shape. It depends only on the potential shape of the considered well: the other well details have no effect on the escape rate. It also depends on the temperature and barrier height only through $\beta\mathcal{B} = \mathcal{B}/(k_B T)$.

When the barrier $\mathcal{B}$ is high ($\beta\mathcal{B} \gg 1$), then the denominator to Eq. 21 can be seen as the average period in the bottom of the well. Since most of the time a quadratic approximation will hold at the thermal noise level, this denominator is simply a period $\bar{\mathcal{T}}^+(0) = \mathcal{T}_0 = 2\pi/\omega_0$ of this harmonic oscillator, and we get a very simple expression matching Kramers' expression:

$$\Gamma^+ = \frac{1}{\mathcal{T}_0} e^{-\beta\mathcal{B}}. \tag{22}$$

Let us now study the switching rate in the double well potential, that is the rate at which the system crosses the barrier, regardless of the direction. For an energy $E > \mathcal{B}$, this will occur twice every $\mathcal{T}(E)$. The switching rate Γ thus writes:

$$\Gamma = \iint_{E>\mathcal{B}} dE\, dt \, \frac{2}{\bar{\mathcal{T}}(E)} P(t, E) \tag{23}$$

$$= \frac{2e^{-\beta\mathcal{B}}}{\int_0^\infty dE \; \beta e^{-\beta E} \bar{\mathcal{T}}(E)} \tag{24}$$

$$= \frac{2}{1/\Gamma^+ + 1/\Gamma^-} \tag{25}$$

For a symmetric double well, the switching rate is equal to the escape rate of a single well: $\Gamma = \Gamma^+ = \Gamma^-$.

To conclude this section, we apply the computation of the barrier crossing rate to the case of the bi-quadratic symmetric double well $U(x)$. For this, we simply report the value of $\mathcal{T}(E)$ from Eq. 15 into Eq. 24. In Fig. 8, we compare the barrier crossing rate computed from this last equation to the one measured on a numerical simulation of a Langevin dynamics in such a potential: the agreement is excellent. The Kramers approximation holds only in the limit of large barriers. It is interesting to note that though the derivation has been done in the Hamiltonian limit ($Q = \infty$), it is still valid even for moderate Q: using numerical simulation, we need to reach $Q = 0.01$ to see a significant deviation to this model.

4.2 *Effective Temperature Due to the Feedback Delay*

Let us introduce the kinetic temperature T of the oscillator defined through the velocity variance: $\sigma_v^2 = \langle v^2 \rangle = k_B T/m$. At equilibrium in a bi-quadratic potential, the kinetic temperature should match the bath temperature T_0 as prescribed by the Boltzmann distribution. When out of equilibrium, we study the steady state and make the hypothesis that the velocity distribution is still described by a Gaussian. Thus

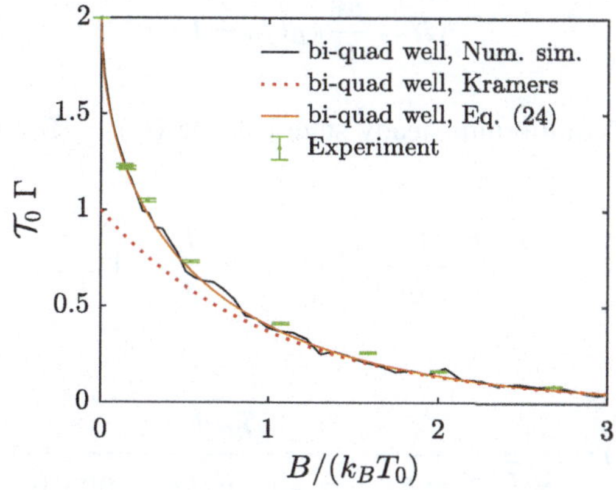

Fig. 8 **Barrier crossing rate** Γ in a bi-quadratic well $U(x) = \mathcal{B}(|x|/x_1 - 1)^2$. The barrier height is $\mathcal{B} = \frac{1}{2}kx_1^2$, and $\mathcal{T}_0 = 2\pi/\omega_0$ is the period in a single harmonic well. At low barrier height, Kramers' simple expression is a factor 2 below the result from Eq. 24. The latter perfectly matches the experimental data or the results from a numerical simulation of a Langevin dynamics in the potential $U(x)$, using a quality factor $Q = 10$

the second moment quantified by T caries all the information needed to describe the statistics of v.

Let us study the non-equilibrium steady state of a wild demon, induced by a total time delay $\tau = \tau_d - \tau_a \neq 0$ between the barrier crossing and the effective potential switch. As anticipated in Sect. 3.1, this delay creates an effective hysteresis $h = |v|\tau$, where v is the velocity of the oscillator when crossing the barrier. The absolute value in this expression stems from the fact that only the velocity sign that matches the barrier crossing is considered (for example positive velocity for upward crossing). The mean value $\langle h \rangle$ can be computed knowing the speed PDF.

$$\langle h \rangle = \langle |v| \rangle\, \tau = \int_0^\infty |v| \frac{e^{-\frac{v^2}{2\sigma_v^2}}}{\sigma_v \sqrt{2\pi}} dv\, \tau \tag{26}$$

$$= \sqrt{\frac{2}{\pi}} \sigma_v \tau = \sqrt{\frac{2}{\pi} \frac{k_B T}{m}}\, \tau \tag{27}$$

The switching rate $\Gamma(\beta\mathcal{B})$ between the 2 wells is given by Eq. 24, and each crossing implies an energy extraction $\Delta U = 2kx_1 h = 2\sqrt{2k\mathcal{B}}h$. The mean power extracted from the oscillator is therefore

$$\langle \dot{U} \rangle = \Gamma \langle \Delta U \rangle = \frac{4}{\sqrt{\pi}} k_B T \sqrt{\frac{\mathcal{B}}{k_B T}}\, \Gamma\left(\frac{\mathcal{B}}{k_B T}\right) \omega_0 \tau. \tag{28}$$

In the steady state, this power is equilibrated by the heat flux coming from the bath, which is expressed within the framework of stochastic thermodynamics as [13, 27]

$$\langle \dot{Q} \rangle = \frac{\omega_0}{Q} k_B (T_0 - T) \tag{29}$$

Writing the out-of-equilibrium steady state balance $\langle \dot{U} \rangle = \langle \dot{Q} \rangle$, we derive the following equation:

$$\gamma \left(\frac{\mathcal{B}}{k_B T_0} \frac{T_0}{T} \right) Q\omega_0 \tau = \frac{T_0}{T} - 1 \tag{30}$$

where

$$\gamma(b) = \frac{4}{\sqrt{\pi}} \frac{\sqrt{b} e^{-b}}{2\pi - \pi e^{-b} + 2\int_1^\infty d\epsilon b e^{-\epsilon b} \sin^{-1}(\epsilon^{-\frac{1}{2}})}. \tag{31}$$

The solution to this equation are a function of $Q\omega_0\tau$, and can be numerically computed, and are plotted in Fig. 9 versus the barrier height $\mathcal{B}$. As expected, the oscillator is warmed up by the demon when $\tau < 0$, and cooled down when $\tau > 0$. The amplitude of the temperature change depends on $\mathcal{B}$. For high $\mathcal{B}$, ΔU is large but barrier crossing are scarce, so the temperature is unchanged. On the contrary, for low barrier heights, crossings are frequent but imply a low energy exchange, leaving the system at the thermostat temperature. For intermediate values $B \sim k_B T_0$, crossing are frequent and induce a sizable energy exchange, thus an effective temperature departing from the T_0.

The function $\gamma(b)$ presents a global maximum $\gamma^* = 0.19$ in $b^* = 0.35$, allowing to compute the extremum temperature and corresponding energy barrier

$$T^* = \frac{T_0}{1 + \gamma^* Q\omega_0\tau} \tag{32}$$

$$\mathcal{B}^* = b^* k_B T^* \tag{33}$$

The extremum temperature is thus a function of $Q\omega_0\tau$. In particular, no kinetic temperature change is expected if $\gamma^* Q\omega_0\tau \ll 1$. Figure 9b demonstrate how this description is pertinent and matches the experimental observations.

Taming the demon thus requires to be able tune the feedback delay to be much smaller than $1/(\gamma^* Q\omega_0)$, a challenge increasing with the quality factor of the oscillator. In our setup ($f_0 = 1270$ Hz), a quality factor $Q = 100$ requires $\tau \ll 6.6\,\mu$s. With an deterministic delay of $\tau_d =\sim 1\,\mu$s on top of the phase delay introduced by the filters, the feedback loop delay necessitate an anticipation of the signal to be transparent in such stringent conditions. To have an effective criterion on the phase of the signal conditioning process, let us fix a maximum heating/cooling to 5% of T_0, that is $\gamma^* Q\omega_0|\tau| < 0.05$. A time delay τ translates into a phase $\phi = \omega\tau$, hence the criterion is simply:

$$\phi < \frac{0.26}{Q} \frac{\omega}{\omega_0}. \tag{34}$$

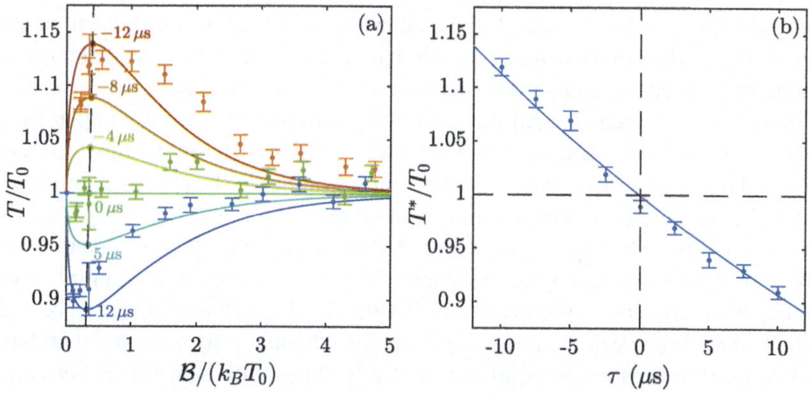

Fig. 9 Consequences of a delay on the kinetic temperature. a T/T_0 is plotted as a function of the barrier height $\mathcal{B}$ between the wells, for various values of the feedback delay τ from $-12\,\mu$s to $+12\,\mu$s. The continuous curves are computed with the model prediction provided by Eq. (30), with $Q\omega_0 = 5.4 \times 10^4$ rad/s. The dependence on $\mathcal{B}$ comes from the balance between the barrier crossing rate and the energy exchanges associated to the delay at each switch. The extremum temperature T^* is well described by Eq. 33 (black dashed line). The experimental data are recorded in air ($Q \sim 10$), with τ_a adjusted to probe three different delays: $\tau = 10\,\mu$s (late trigger), $\tau = 0\,\mu$s (behaving demon), and $\tau = -10\,\mu$s (early trigger). When $\tau = 0$, the kinetic temperature deviates at most by 3% from the thermostat temperature. (b) Extremum temperature T^* as a function of the feedback delay τ, described by Eq. 32. Positive delay τ cools the system down, while negative ones warms the system up. The experimental data corresponds to a measurement at $\mathcal{B} = b^* k_B T_0$, and is very close to the prediction

This criterion is reported in Fig. 5. We see that we manage to fulfill the criterion in a reasonable frequency range around the resonance, where most of energy of the motion is concentrated. All in all, a well tuned feedback allows the kinetic temperature to match the thermostat one within 3%. Moreover, the demon can also be unleashed on purpose by tuning the anticipation delay τ_a so as to create a non-equilibrium exchange of energy with the thermostat.

5 Conclusions

In this chapter, we describe how a virtual potential can be realized using a feedback loop to modify the dynamics of an underdamped system, in our case is a micro-mechanical oscillator. This technique produces a tunable non linear virtual potential which is indistinguishable from a physical one on the example demonstrated here: a bi-quadratic double well. To get such a result we take into account and overcome the drawbacks of the method, which are induced for example by the feedback delay and by noise in the commutation threshold between the two wells. We discuss in details how to fine tune the feedback parameters in order to get a "perfect" virtual-potential and we show how the delay (anticipation) of the feedback introduces a cooling (heating) of the device. As an example of application, we present a detailed experimental and theoretical analysis of the jumping rate between two potential wells showing the correction to Kramer's rate for small energy barriers.

The use of non linear virtual-potential in underdamped oscillators open the way to many new and interesting applications which present different features from those performed in overdamped systems. For example, our device has been recently used to measure Landauer's bound in underdamped systems [12, 13, 25, 27, 28], underlying their advantages with respect to overdamped ones. Among other applications is the analysis of the energetics of the feedback: the device can be used as an information engine in which the information acquired by the feedback (the demon) can be used to produce work. We overlooked this aspect in this chapter, only showing that demons could be used to create non-equilibrium steady states with heat fluxes between the oscillator and the thermostat. The analysis of the demon from the information prism is still under study and will be the subject of future works.

Competing Interests The authors have no conflicts of interest to declare that are relevant to the content of this chapter.

Acknowledgements This work has been financially supported by the Agence Nationale de la Recherche through grant ANR-18-CE30-0013 and by the FQXi Foundation, Grant No. FQXi-IAF19-05, "Information as a fuel in colloids and super-conducting quantum circuits".

References

1. Gavrilov, M., Jun, Y., Bechhoefer, J.: Real-time calibration of a feedback trap. Rev. Sci. Instrum. **85**, 095102 (2014). https://doi.org/10.1063/1.4894383
2. Gavrilov, M., Bechhoefer, J.: Arbitrarily slow, non-quasistatic, isothermal transformations. EPL Europhys. Lett. **114**, 50002 (2016). https://doi.org/10.1209/0295-5075/114/50002
3. Adam, E.: Cohen, Control of nanoparticles with arbitrary two-dimensional force fields. Phys. Rev. Lett. **94**, 118102 (2005). https://doi.org/10.1103/PhysRevLett.94.118102
4. Jun, Y., Bechhoefer, J.: Virtual potentials for feedback traps. Phys. Rev. E **86**, 061106 (2012). https://doi.org/10.1103/PhysRevE.86.061106
5. Albay, J.A.C., Lai, P.-Y., Jun, Y.: Realization of finite-rate isothermal compression and expansion using optical feedback trap. Appl. Phys. Lett. **116**, 103706 (2020). https://doi.org/10.1063/1.5143602
6. Bérut, A., Arakelyan, A., Petrosyan, A., Ciliberto, S., Dillenschneider, R., Lutz, E.: Experimental verification of Landauer's principle linking information and thermodynamics. Nature **483**, 187 (2012). https://doi.org/10.1038/nature10872
7. Bérut, A., Petrosyan, A., Ciliberto, S.: Information and thermodynamics: experimental verification of Landauer's erasure principle. J. Stat. Mech. **2015**, P06015 (2015). https://doi.org/10.1088/1742-5468/2015/06/p06015
8. Hong, J., Lambson, B., Dhuey, S., Bokor, J.: Experimental test of Landauer's principle in single-bit operations on nanomagnetic memory bits. Sci. Adv. **2**, e1501492 (2016). https://doi.org/10.1126/sciadv.1501492
9. Martini, L., Pancaldi, M., Madami, M., Vavassori, P., Gubbiotti, G., Tacchi, S., Hartmann, F., Emmerling, M., Höfling, S., Worschech, L., Carlotti, G.: Experimental and theoretical analysis of Landauer erasure in nano-magnetic switches of different sizes. Nano Energy **19**, 108 (2016). https://doi.org/10.1016/j.nanoen.2015.10.028

10. Proesmans, K., Ehrich, J., Bechhoefer, J.: Finite-time Landauer principle. Phys. Rev. Lett. **125**, 100602 (2020). https://doi.org/10.1103/PhysRevLett.125.100602
11. Jun, Y., Gavrilov, M., Bechhoefer, J.: High-precision test of Landauer's principle in a feedback trap. Phys. Rev. Lett. **113**, 190601 (2014). https://doi.org/10.1103/PhysRevLett.113.190601
12. Dago, S., Pereda, J., Barros, N., Ciliberto, S., Bellon, L.: Information and thermodynamics: fast and precise approach to Landauer's bound in an underdamped micromechanical oscillator. Phys. Rev. Lett. **126**, 170601 (2021). https://doi.org/10.1103/PhysRevLett.126.170601
13. Dago, S., Bellon, L.: Dynamics of information erasure and extension of Landauer's bound to fast processes. Phys. Rev. Lett. **128**, 070604 (2022). https://doi.org/10.1103/PhysRevLett.128.070604
14. Gieseler, J., Millen, J.: Levitated nanoparticles for microscopic thermodynamics–a review. Entropy **20**, 326 (2018). https://doi.org/10.3390/e20050326
15. Gieseler, J., Novotny, L., Moritz, C., Dellago, C.: Non-equilibrium steady state of a driven levitated particle with feedback cooling. New J. Phys. **17**, 045011 (2015). https://doi.org/10.1088/1367-2630/17/4/045011
16. Seifert, U.: Stochastic thermodynamics, fluctuation theorems and molecular machines. Rep. Prog. Phys. **75**, 126001 (2012). https://doi.org/10.1088/0034-4885/75/12/126001
17. Kyung Hyuk Kim and Hong Qian: Entropy production of Brownian macromolecules with inertia. Phys. Rev. Lett. **93**, 120602 (2004). https://doi.org/10.1103/PhysRevLett.93.120602
18. Granger, L., Kantz, H.: Thermodynamic cost of measurements. Phys. Rev. E **84**, 061110 (2011). https://doi.org/10.1103/PhysRevE.84.061110
19. Rosinberg, M.L., Munakata, T., Tarjus, G.: Stochastic thermodynamics of Langevin systems under time-delayed feedback control: second-law-like inequalities. Phys. Rev. E **91**, 042114 (2015). https://doi.org/10.1103/PhysRevE.91.042114
20. Sagawa, T., Ueda, M.: Generalized Jarzynski equality under nonequilibrium feedback control. Phys. Rev. Lett. **104**, 090602 (2010). https://doi.org/10.1103/PhysRevLett.104.090602
21. Lahiri, S., Rana, S., Jayannavar, A.M.: Fluctuation theorems in the presence of information gain and feedback. J. Phys. A: Math. Theor. **45**, 065002 (2012). https://doi.org/10.1088/1751-8113/45/6/065002
22. Paolino, P., Aguilar Sandoval, F., Bellon, L.: Quadrature phase interferometer for high resolution force spectroscopy. Rev. Sci. Instrum. **84**, 095001 (2013). https://doi.org/10.1063/1.4819743
23. John Elie Sader: Frequency response of cantilever beams immersed in viscous fluids with applications to the atomic force microscope. J. Appl. Phys. **84**, 64 (1998). https://doi.org/10.1063/1.368002
24. Bellon, L.: Thermal noise of micro cantilevers in viscous fluids. J. Appl. Phys. **104**, 104906 (2008). https://doi.org/10.1063/1.3021102
25. Dago, S., Ciliberto, S., Bellon, L.: Adiabatic computing for optimal thermodynamic efficiency of information processing. Proc. Nat. Acad. Sci. **120**, e2301742120 (2023). https://doi.org/10.1073/pnas.2301742120
26. Butt, H.-J., Cappella, B., Kappl, M.: Force measurements with the atomic force microscope: technique, interpretation and applications. Surf. Sci. Rep. **59**, 1 (2005). https://doi.org/10.1016/j.surfrep.2005.08.003
27. Dago, S., Pereda, J., Ciliberto, S., Bellon, L.: Virtual double-well potential for an underdamped oscillator created by a feedback loop. J. Stat. Mech. **2022**, 053209 (2022). https://doi.org/10.1088/1742-5468/ac6d62
28. Dago, S., Ciliberto, S., Bellon, L.: Reliability and operation cost of underdamped memories during cyclic erasures. Adv. Phys. Res. 2300074 (2023). https://doi.org/10.1002/apxr.202300074

Symmetry, Chirality and Unidirectional Motion

Jean-Pierre Launay

Dedicated to Prof. David Carillo, Universidad Catolica de Valparaiso, Chile, for his 85th birthday.

Abstract Several aspects of chirality are presented: Curie principle, types of chirality, atropoisomerism, true and false chirality. Then, considering the quantum mechanical aspects, we discuss Hund's paradox–namely why, although the complete Hamiltonian of a molecule is parity-invariant, broken symmetry enantiomeric states appear and can be frequently isolated as macroscopic samples. A first approach is based on the existence of a potential energy curve with two minima, which can be rationalized in the frame of Pseudo Jahn Teller effect. But a more complete description is now available through the environment induced decoherence of superposed states. Coming back to the standard (semi-classical) Born-Oppenheimer description based on potential energy surfaces, we analyze qualitatively a simple example, 2,2'-dibromobiphenyl to point out the interaction between different degrees of freedom. This is the starting point for a rational construction of a system with built-in symmetry allowing unidirectional motion, like Feringa's molecule. The role of chirality as a necessary ingredient is stressed, as well as the occurrence of conical intersections in the photochemical excitation step. Then, "flashing ratchet" mechanisms, involving a potential energy curve with sawtooth asymmetric profile, are introduced. Experimental examples of molecular systems deposited on surfaces and excited by the tunneling current of a STM are discussed.

J.-P. Launay (✉)
Centre d'élaboration de matériaux et d'études structurales (CEMES). Centre national de la recherche scientifique (CNRS), Université de Toulouse, 29 rue J. Marvig, BP 4347, F-31055 Toulouse Cedex, France
e-mail: jean-pierre.launay@cemes.fr

X. Bouju and C. Joachim (eds.), *Crossroad of Maxwell Demon*, Advances in Atom and Single Molecule Machines, https://doi.org/10.1007/978-3-031-57904-2_7

1 Introduction

Unidirectional motion is one of the great challenges of Nanosciences. It is of course very familiar to us in everyday's life, but its achievement at the nanoscale necessitates the combination of several modes of reasoning. When facing a complex nanoscale object, our instinctive way of reasoning is frequently biased by our tendency to attribute an intention to this inanimate object, without properly taking into account the importance and ubiquity of random motions, which generally lead to unexpected behaviours. Symmetry, including chirality or a- chirality, plays of course a role as in the macroscopic world, as recognized by Pierre Curie as early as 1894 [1]. But there are more subtle concepts to take into account, like microreversibility [2]. This principle states that in a molecular transformation, the forward and backward processes must occur along the same path and involve the same intermediates. Thus hysteresis effects, so common in the macroscopic world, are not possible.

Unidirectional motion necessitates some symmetry breakdown to discriminate forward/backward evolutions. In this Chapter we consider only rotation motions, and thus the distinction between clockwise/anticlockwise motions. For this purpose, we'll make use of potential energy curves or surfaces with a rotation angle as one of the nuclear coordinate, and they will be shown to be intimately linked to the property of chirality.

In Sect. 2, we'll remind the importance of symmetry, coming back to the earlier Curie Principle. Since all examples treated here are molecules, we'll recall some common cases of chirality, including the particular case of atropoisomerism. (This will be only a short résumé of what can be found in usual chemistry textbooks.)

Section 3 will be truly at the interface between Chemistry and Quantum Mechanics, by considering the famous Hund's Paradox (1927). This puzzling observation is that, although the Hamiltonian of a molecule is right/left symmetrical, so that we could expect solutions of the Hamiltonian to present right/ left symmetry, it is possible to isolate chiral samples (this is in particular crucial in the pharmaceutical industry). Progresses in the resolution of this paradox will be evoked, in particular through the theory of environment induced decoherence.

In Sect. 4, we'll consider examples of molecules with an internal twisting degree of freedom. The goal of this section will be to start from the simplest chemical structure, a substituted biphenyl, and by increasing progressively the complexity, to make the link with the successful example by Feringa of a molecule able to perform a unidirectional internal rotation upon a suitable sequence of photochemical and thermal excitations. The role of chirality will be highlighted, as well as the occurrence of conical intersections skewing the photochemical process in a definite direction.

Section 5 will be devoted to the flashing ratchets mechanisms. As they seem to play a role in the behavior of molecules deposited on a surface and excited by tunneling from a STM tip, several experimental examples will be discussed, showing the relation between the chirality of the structure and the unidirectional rotation.

Finally, in Sect. 6, some conclusions will be presented with a short discussion about the puzzling obtaining of symmetrical energy profiles in theoretical calculations, even for a molecule with a clearly asymmetric structure.

2 Symmetry and Chirality

Since Pierre Curie statement [1], it is admitted that the symmetry of effects is related in the symmetry of causes. The symmetry of causes encompasses the symmetry of the system (in our case its three-dimensional structure but also the symmetry of external actions, for instance the position of an external electric or magnetic field, or the precise position of an STM tip). Although, deviations from this general principle can occur, it constitutes a powerful basis for a first analysis of a process.

A nice example of the role of symmetry is given by the behavior of the "rattleback" (Fig. 1). This rodlike object with the general shape of a canoe exhibits a strange behaviour: it rotates smoothly only in one direction (for instance clockwise) even when spun in the reverse direction. This effect is ultimately linked to a subtle asymmetry in the structure, and the detailed explanation can be found in [3].

In this context of structure-behaviour relation and since we are concerned with unidirectional rotation motion, the key structural factor is chirality, that is symmetry with respect to parity (mirror reflection). Chirality is well documented in chemistry, and we recall here only a few typical examples.

Point chirality. It can be illustrated by the well-known case of asymmetric carbon, that is a carbon atom attached to four different types of atoms or groups of atoms (Fig. 2a). To define the exact geometry, one uses the Cahn-Ingold-Prelog priority rules for ranking the four substituents [4]a. The resulting denomination is either R (rectus) or S (sinister) according to the direction of the arrangement of three substituents. However, in the following of this chapter, we'll use the R/L denomination, more widely used in physics. The other types of chirality listed below do not involve a chiral center.

Axial chirality. In this case, chirality arises from the disposition of different substituents around a specific axis. Typical examples are substituted biphenyls, and

Fig. 1 A rattleback

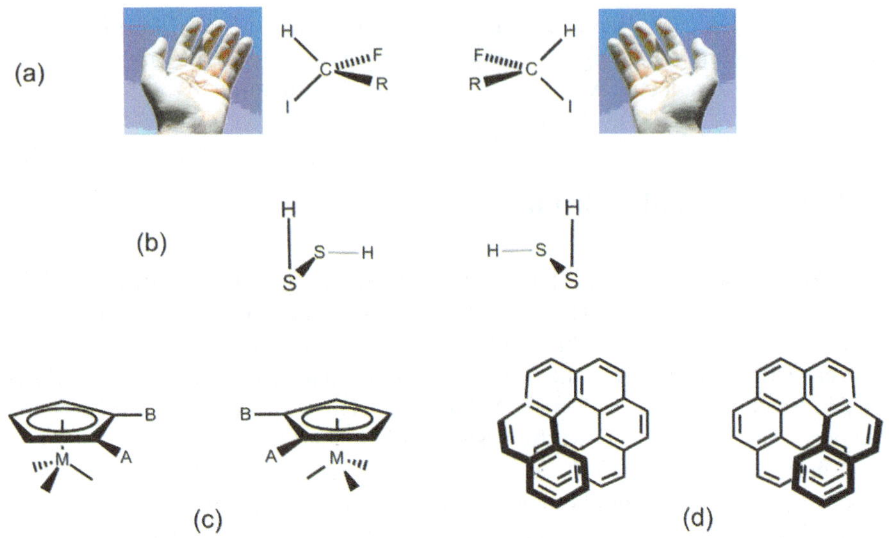

Fig. 2 Some different types of chirality in chemical structures. **a** Point chirality (asymmetric carbon). **b** Axial chirality (H_2S_2). **c** Planar chirality. **d** Intrinsic chirality (helicene)

we'll treat also the simpler case of H_2X_2 molecules (X=O or S) in their *gauche* conformation (Fig. 2b).They will be discussed extensively in Sect. 3.

Planar chirality. Here we can define a plane containing as much atoms as possible, but not all, and chirality arises from the disposition of substituents in the plane and out of the plane (Fig. 2c). Typical examples are provided by metal coordination complexes containing a cyclopentadienyl ring and various substituents in a piano stool geometry.

Intrinsic chirality. Examples of intrinsic chirality are provided by helicenes (Fig. 2d). The chirality is qualified as intrinsic, because one cannot define neither a center, nor an axis or plane around which a specific arrangement of substituents could be defined. Instead, the chirality is due to the overall shape of the molecule. In the example of Fig. 2d, the molecule is made of seven phenyl rings whose tendency to form a planar arrangement is hampered by molecular overcrowding. The resulting shape is reminiscent of a helix or corkscrew, with two possible rotation directions. This type of chirality will play a role in Sect. 4 (Feringa).

Atropoisomerism. A special case is represented by *atropoisomerism*. The term is built from the greek "a" and "tropos" (turn), meaning "without turn" [4]b. It results from hindered rotation around one or several single bonds and this can give rise to axial chirality. If the rotation barrier can be overcome by thermal excitation, at high temperature, there is free rotation, thus no distinct isomers, but at low temperature, the geometry is quenched and several atropoisomers can be identified, with however the possibility of interconversion. An arbitrary criterion for the existence of atropoisomers at room temperature is a rotation barrier above 22 kcal/mol, corresponding

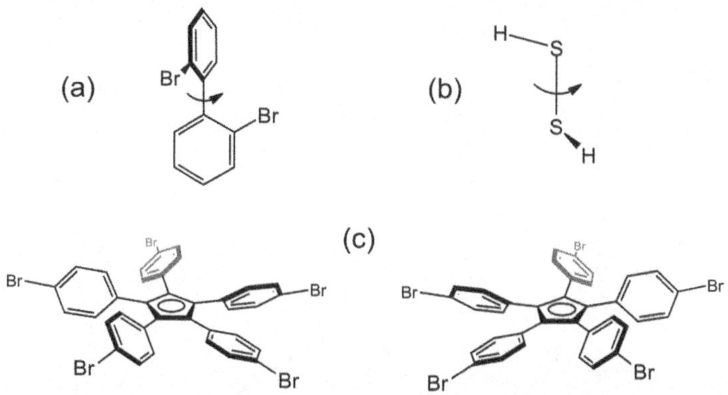

Fig. 3 Some examples of atropoisomerism. **a** 2,2'-dibromobiphenyl. **b** H_2S_2. **c** penta (bromophenyl) cyclopentadienyl with correlated tilting of the 5 phenyl rings (2 enantiomers shown)

to a half-life above 1000 s. Atropoisomerism will be encountered below in the case of 2,2'-dibromobiphenyl (Fig. 3a). It manifests also in the case of H_2S_2 but at very low temperatures, since the barrier height is only 4 kcal/ mol (Fig. 3b). Finally there are more complex cases, for instance in pentaphenyl-cyclopentadienyl anion : the five phenyl groups cannot orient independently and there are thus only two atropoisomers (Fig. 3c), the phenyl groups adopting the disposition of propeller blades with opposite chiralities [5].

True and false chirality. Following Barron [6], it appears necessary in some physics experiments involving magnetic fields or motion, to distinguish between true and false chirality. This author introduced the following definition: "*true chirality is possessed by systems that exist in two distinct enantiomeric states that are interconverted by space inversion but not by time reversal combined with any proper spatial rotation*". Note that "space inversion" $(x, y, z \rightarrow -x, -y, -z)$ can be replaced by "reflection in a plane followed by a 180° rotation about an axis perpendicular to the plane", and this operation is called Parity in the following. The subtlety of the definition with its relevance to time inversion becomes important when motion is involved. Thus in Fig. 4a, two spinning cones seem to be enantiomers because they correspond to each other by the P operation. However, the correspondence could be *also* achieved by time reversal T (reversing the rotation direction) and performing a rotation by π, and thus the two cones are not truly chiral. By contrast, in Fig. 4b where the cones are moving, they are related by the parity operator P, but cannot be superposed by a time reversal T followed by a π rotation. Here the cones *are* truly chiral.

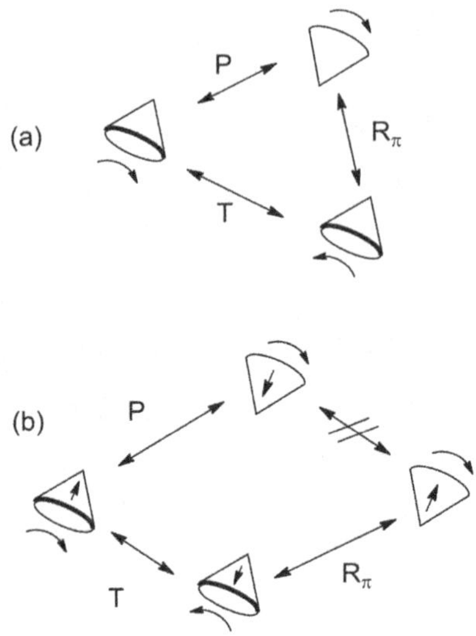

Fig. 4 True and false chirality illustrated with rotating cones. **a** The Parity inversion on one hand and the Time Inversion (reverting the rotation direction) followed by a 180° rotation on the other hand generate the same enantiomeric state. This is *false chirality*. **b** Parity inversion and the sequence of Time Inversion and 180° rotation generate different enantiomers. This is *true chirality*. Adapted with the permission from Ref. [6] Copyright 2018, American Physical Society

3 Hund's Paradox (1927)

As early as 1927, F. Hund formulated the following paradox [7]: how is it possible to observe and often separate enantiomers of molecules, while the molecular Hamiltonian for a collection of electrons and nuclei is parity-invariant? (it does not contain angles, but only interparticle distances). Taking $|R\rangle$ and $|L\rangle$ as basis states, it means that the quantum states superpositions like $2^{-1/2}(|R\rangle \pm |L\rangle)$ or more generally $2^{-1/2}(c_L|L\rangle + c_R e^{i\phi}|R\rangle)$ with ϕ an arbitrary phase angle, although they are *bona fide* eigenfunctions are not observed. Note that most chemists are not puzzled by this situation because they visualize easily the 3D structure of $|R\rangle$ and $|L\rangle$ and can check the orthodox disposition of atoms and bonds, while the quantum state superposition looks as some kind of Schrödinger's cat. To make things more concrete, let us use the case of H_2S_2 (Fig. 5) to find elements of answer to the paradox.

Note that in the following we shall *not* consider the effect of Parity Violation (the small energy difference between R and L enantiomers due to the weak nuclear force which is chirality sensitive). This effect is indeed extremely small, of the order of 10^{-16} eV [8], well below the energies of physical and chemical processes involved here.

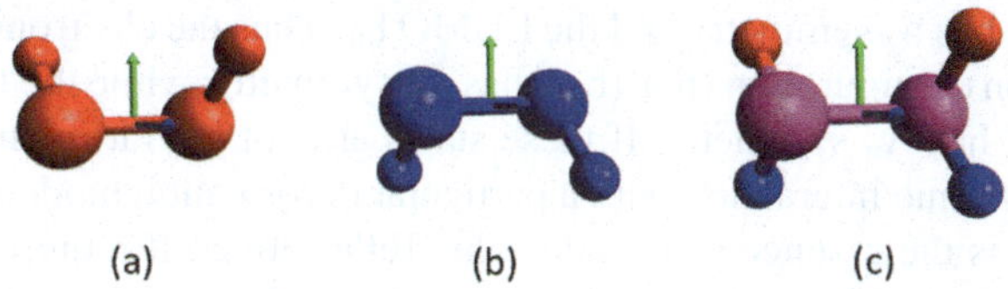

Fig. 5 **a b** Chiral structures of H_2S_2. **c** An artist view of a quantum state superposition structure. The two R and L components are oriented to share the same C_2 axis (vertical, in green) and are related by an inversion center. The magenta spheres encompass two close positions for the S atoms belonging to the R and L components

3.1 Attempt to Explain the Paradox by the Potential Energy Curve

Optical isomers correspond to minima in some potential energy curve, for H_2S_2 the twisting around the central S–S bond. In this way of reasoning we implicitly make use of the Born-Oppenheimer approximation and assume that it is possible to compute the molecular energy as a function of a deformation parameter. A powerful tool to rationalize the lack of symmetry is the Pseudo-Jahn-Teller formalism (PJTE) [9]. In the following, we use the reasoning developed for the isoelectronic H_2O_2 molecule.

Let us start from an achiral geometry, for instance the trans-planar one (C_{2h} point group). It is consistent with the plausible bond orders and minimizes the steric interaction between H atoms, but is it stable with respect to twisting around the S–S bond? It is easy to compute the shape and properties of the HOMO and LUMO, shown on Fig. 6, and a simple reasoning shows if there is a possibility of instability with respect to the planar geometry.

Fig. 6 **a** HOMO and **b** LUMO of H_2S_2 in the trans-planar state; **c** the q_{au} coordinate

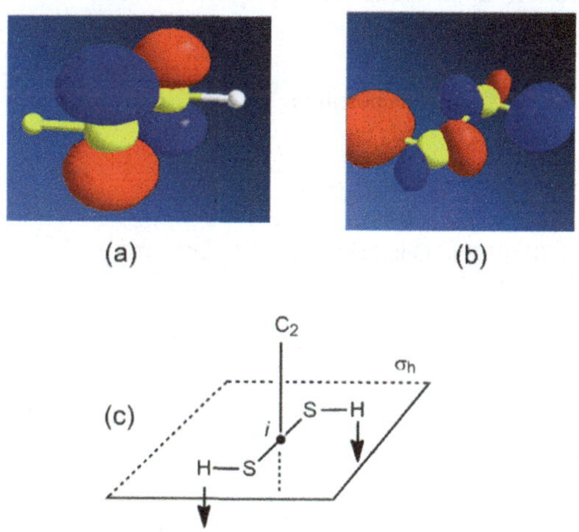

The HOMO has b_g symmetry and the LUMO b_u. Thus the electronic ground state Ψ_0 with electron configuration $(b_g)^2(b_u)^0$ has A_g symmetry, while the Ψ_1 first excited state $(b_g)^1(b_u)^1$ has A_u symmetry. If these states are not too far in energy, they can be mixed by vibronic interaction with a particular deformation mode, stabilizing the structure. This is the essence of Pseudo-Jahn-Teller effect. It requires that a deformation mode exists with the same symmetry as the $\Psi_0\Psi_1$ product, i.e. $A_g \times A_u = A_u$. In the present case, there is indeed such a motion with a_u symmetry favoring a chiral (skewed) geometry (Fig. 6) [10]. But it competes with the contribution of the ground state itself which has the opposite effect. The deformation can win if the excited state is not too high in energy.

Extensive high-level calculations at the ab initio or DFT level have confirmed a non-planar (chiral) structure with *gauche* geometry and a dihedral angle close to 90° in the case of H_2O_2 and H_2S_2, in agreement with experiment [10–12]. These calculations provide also the height of the energy barriers for rotation. Although high-level ab initio or DFT calculations predict correctly the geometry, they lack the explanatory and predictive power of PJTE.

Thus, the potential energy as a function of twisting angle exhibits two minima, but these are not eigenstates because they have no symmetry type (they are neither symmetric nor antisymmetric) with respect to mirror inversion. As seen above, the true eigenstates can be built by combining $|R\rangle$ and $|L\rangle$ to generate for instance symmetric or antisymmetric wavefunctions.

For such two level systems, it is well known that there is a possibility of nuclear tunneling if the system is prepared at some time in the $|R\rangle$ or $|L\rangle$ state (see Fig. 7). Then the system should oscillate periodically between the $|R\rangle$ and $|L\rangle$ states with a period $h/\Delta E$ where ΔE is the (very small) difference in energy between the lowest eigenstates (Fig. 7). For H_2S_2, the ΔE splitting is 5×10^{-6} cm^{-1}, giving a period of 6×10^{-6} s [12].

The tunneling process is consistent with the fact that neither $|R\rangle$ or $|L\rangle$ are eigenfunctions of the molecular Hamiltonian, but it does not explain why the system does

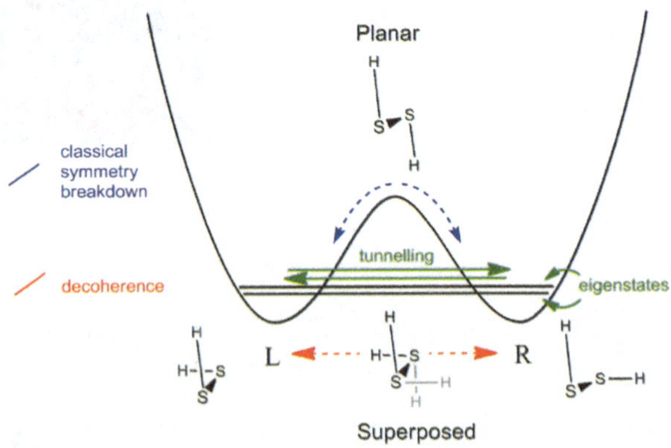

Fig. 7 Two-wells potential energy curve of H_2S_2, showing the tunnel effect through the barrier and the different discussed structures (L, R, Planar, Quantum state Superposition)

not spontaneously fall and stay in the true ground state $2^{-1/2}(|R\rangle + |L\rangle)$, and why it is difficult to observe periodic tunneling.

3.2 Explanation by Decoherence

The concept of environmental decoherence appeared at the beginning of the 80's [13] and seems adapted to resolve the paradox. Decoherence is the process by which a quantum system in a superposition state is brought in a given particular component state (for instance $|R\rangle$) by interaction with the environment. At first sight it is a way to escape and ignore the famous and hard-to-believe process of wavefunction collapse which should occur at the precise moment of measurement, according to the orthodox formulation of quantum mechanics. Environmental decoherence is achieved by interaction of the quantum object with a huge number of neighbouring particles or even photons, and it occurs before the intervention of an observer and anyway independently of its presence or state of consciousness.

Environmental decoherence is a dynamic process particularly difficult to modelize. But the challenge was met by Trost and Hornberger in 2009, through a detailed computer simulation of the behavior of a D_2S_2 molecule bombarded by He atoms [14] (Fig. 8).

The calculation uses a scattering matrix with distinct scattering amplitudes for the R and L components (same amplitude but a different phase). The interaction between D_2S_2 and He is of the London type, that is between induced dipoles and even induced quadrupole for D_2S_2. A large number of possible orientations are considered and there is an averaging over rotational states. This represented about 1000 coupled equations to solve simultaneously. Finally, the simulation was performed for temperatures from to 0 to 10 K, and the results extrapolated beyond. The computation time was about one month.

The output was obtained as scattering cross sections as functions of temperature. The decoherence cross section represented about 10 % of total cross sections at 10 K, and about 25 % at 300 K, which is considered as very efficient. The decoherence rate was found much larger than the tunneling rate. The chiral states appear to disappear at a still lower rate, thus showing that the decoherence process is irreversible. From

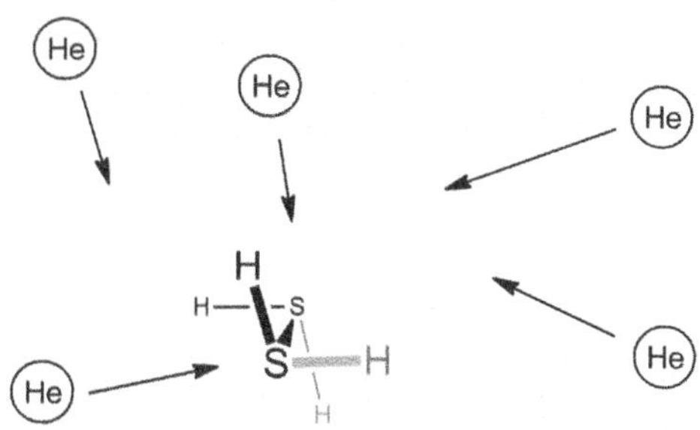

Fig. 8 Scheme of the scattering between a He particle and H_2S_2 in the quantum state superposition

a practical point view, the critical pressure for a Helium atmosphere to achieve decoherence was found as *ca* 10^{-5} mbar at 300 K. An interesting observation was made when looking at the different components of the London interaction. They are described by polarizability tensors, and an important contribution comes from the quadrupole-dipole polarizability, which is *chirally sensitive*. This particular term describes the induction of a quadrupolar moment by an electric field, or conversely the induction of a dipole by a field gradient [15]. A pictorial illustration is proposed in the twisted arc model of chiral molecules [16] showing that this quantity is dependent on H_2S_2 enantiomer for a given geometry of the collision. Thus the corresponding interaction has a different effect on the two components of a R/L superposition and acts as a "kick" changing the relative phase of the two parts of the wavefunction [17].

A simpler explanation is provided by Bahrami and Bassi, considering D_2S_2 and He as made of hard spheres. They argue that the important interaction is between He and the D atoms (in quantum state superposition) and this necessitates that He arrives from a direction which is not shielded by sulfur atoms. Then, the geometry of the collision predicts the possibility and efficiency of decoherence [18].

Returning to the numerical simulation, it shows the decoherence of the quantum state superposition at the expense of one of the chiral states, and in addition that the chiral states are robust to further evolution, so that the process is irreversible. But why this difference in the behavior of superposition and chiral states? In fact, not all states or superpositions are equal with respect to the interaction with environment. Since the beginning of decoherence theory, it was asserted that among the vast number of possible states, some of them (called pointer states) are more adapted to experimental detection. The detailed theory requires the description of the system, the measurement device, *and the environment* [13] (while the historical formulation of quantum mechanics did not consider the last element). In this spirit, chiral states are pointer states, but superpositions are not. Why pointer states present such a privilege? A hint is provided by the more recent development of decoherence, Quantum Darwinism [19]. Pointer states interact with their environment by modifying it (for instance modifying the position or velocity of a colliding particle, or scattering a photon), but for these states, the interaction is particularly strong, and they have the possibility to generate a large number of copies of their influence. Then, by a process analogous to natural selection, they are more easily detected than the other states, because *in fine* many independent observers can make a measurement and agree on the outcome [19]. A final question is how to identify from their structure the best candidates to the status of pointer states? The mathematical study of Quantum Brownian Motion by means of the time evolution of the density matrix of the complete system shows that "only relatively localized states survive interaction with the environment". At variance, "spread-out superpositions of localized wave packets are rapidly destroyed" [20].

4 Trekking on Energy Surfaces–From Biphenyl to Feringa's Molecules

In this section we discuss the case of molecules presenting an intramolecular twisting degree of freedom, in order to identify the minimal structure allowing unidirectional rotation. We come back to the usual approach, based on Born-Oppenheimer approximation and potential energy surfaces as functions of nuclear coordinates. This is a semi-classical model, familiar to chemists and physicochemists, which is pertinent here because we do not consider small energy differences and low temperature experiments as in the previous section. The exploration of the potential energy surfaces will be discussed in analogy with trekking on a rugged landscape with summits, ridges, valley, passes and so on, but in addition upper surfaces corresponding to excited states will play a role. These excited state surfaces will be necessary to overpass some high elevation obstacles and escape microscopic reversibility.

We start from a simple example 2,2'-dibromobiphenyl in which two substituted phenyl groups are linked by an essentially single bond. The rotation around the single bond should be almost free (barrier about 2–3 kcal/mol for biphenyl) but the presence of bulky Br atoms introduces a steric constraint of the order of 20 kcal/mol [21]. We schematize the potential energy surface using two twisting angles, θ between the phenyl groups and β corresponding to the possibility for Br to deviate from the phenyl plane when there is too much strain (Fig. 9).

Figure 9 schematizes the shape of the potential energy surface. (It is inspired from [22], but contrarily to this work we assume here that both Br atoms are free to deviate). For $\beta = 0$ (rigid bromophenyl groups) one can expect "valleys" for $\theta = 90$ or $270°$ (no steric interaction), a high elevation "butte" for $\theta = 0°$ (strong hindrance between the two Br atoms), and a lower one for $\theta = 180°$ (weaker interaction between Br and H atoms) (Fig. 9). If now, Br atoms are allowed to deviate from their respective planes the buttes and valleys appear tilted with respect to the axes, because of the correlation between θ and β: to reduce the steric strain, one can play either on θ or β or a suitable combination of both.

We now consider the case of overcrowded alkenes, where two cycles are connected by a double bond (Fig. 10, in which we have introduced only one Br atom). Then there is a strong electronic preference for planar geometries $\theta = 0$ or $180°$ because of the alkene bond which in addition is conjugated with the double bonds of the cycles. But steric repulsion makes planarity impossible. For instance in octachlorofulvalene, the twisting angle between cycles is 37° [23]. So here we expect some buttes at 0 and 180°, but the main change with respect to dibromobiphenyl is a very high mountain ridge for 90 or 270°. For these θ values, the lowest energy is obtained when Br is relaxed, that is in the plane of its cycle ($\beta = 0$), thus defining a pass. With respect to Fig. 9, the central valley is now cut in two basins of same depth, denoted 1 and 2, and reproduced by symmetry as 3 and 4. If we imagine a trek towards the right starting from 1, there is a huge obstacle to arrive in 2, the $\theta = 90°$ ridge, even through the pass. (As shown below, it can be crossed only by photochemical excitation). Going from 2 to 3 is easier by getting around the butte by any side.

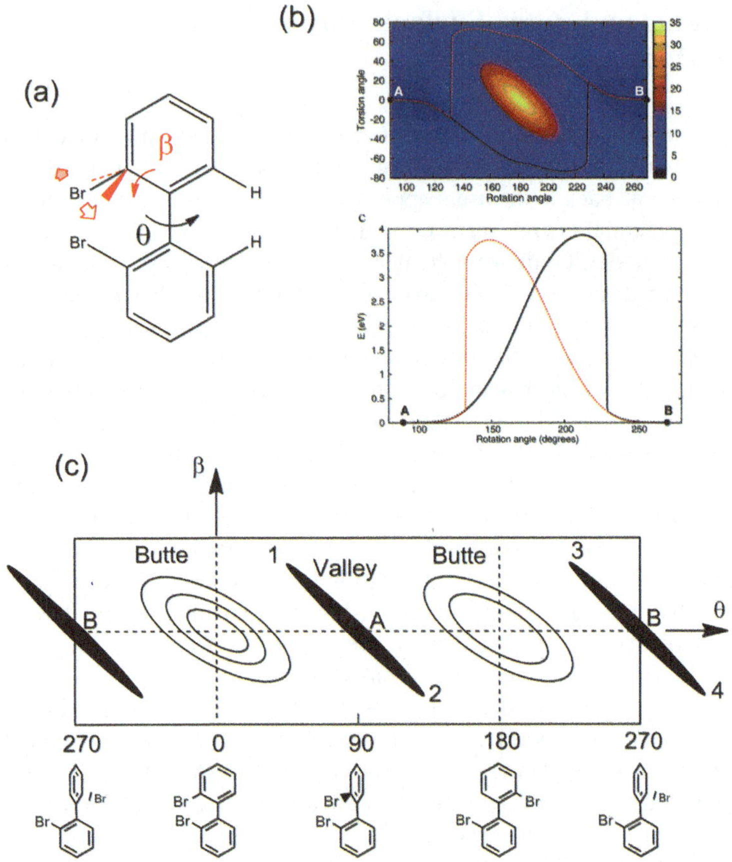

Fig. 9 Predicted potential energy surface for 2,2'-dibromobiphenyl. **a** Internal coordinates, θ, twisting angle between the phenyl rings, θ, twist angle corresponding to deviation of Br with respect to its phenyl plane. **b** above: energy as function of θ (horizontal) and β (vertical); below: energy profile as function of θ with two possible trajectories; points A and B correspond to perpendicular (90 or 270°) geometries. **c** Guessed potential energy surface = $f(\theta, \beta)$ for 2,2'-dibromobiphenyl. Numbers 1–4 at the extremities of the valleys in black will be useful for comparison with Fig. 10. Part (b) reproduced with permission of Royal Society of Chemistry from Echeverria et al., *Nanoscale* 2014 [22]; permission conveyed through Copyright Clearance Center, Inc

The landscape has now gained in complexity but this is not enough to favor a unidirectional motion. If we change θ into $-\theta$ and β into $-\beta$, that is rotating Fig. 10b by 180°, the landscape is exactly the same, so that clockwise and anticlockwise rotations are equally probable, in agreement with microscopic reversibility. One of the last necessary ingredients is chirality (Fig. 11). Adding an asymmetric carbon on the top cycle suppresses its planar structure and discriminates structures symmetrically located on either side of $\theta = 0$ or 180°. Now, basins 1 and 2 or 2 and 3 no longer

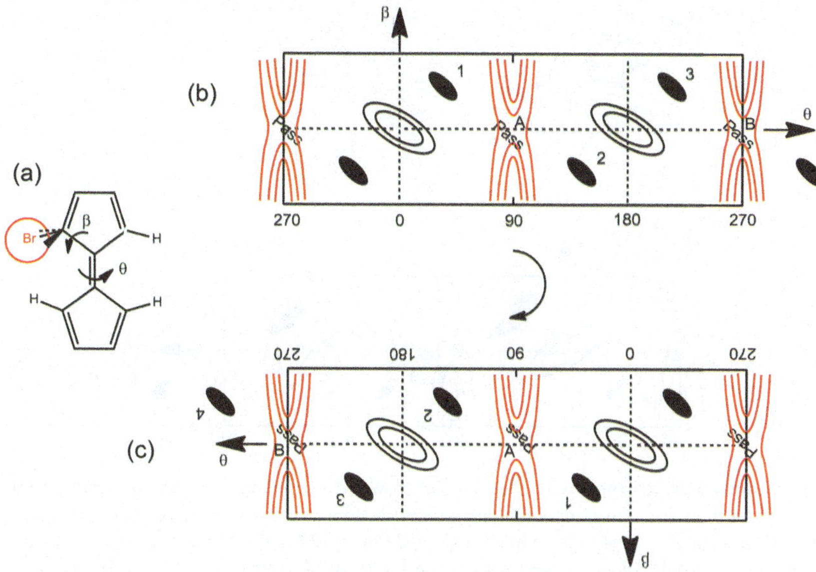

Fig. 10 An example of overcrowded alkene: hypothetical bromo-fulvalene. **a** Structure with a Br close to the double bond. **b** Guessed potential energy surface = $f(\theta, \beta)$. The highest ridge is in red (see text for discussion). **c** Rotation of the diagram by 180° showing that an evolution by changing simultaneously the signs of θ and β leads to an equivalent situation

have the same depth, and a unidirectional motion is possible by a suitable sequence of excitations. This is the structural basis of the behavior of Feringa's molecules.

Figure 12 represents the energy as a function of reaction coordinate for a typical Feringa molecule [24]. Note that the reaction coordinate is no longer a pure twisting mode but a combination of motions, in which θ is dominant and increases from left to right. We recognize the lower energy geometries 1, 2 and 3, with 2 higher than the others, and the very high barrier between 1 and 2, which is of the order of several eV.

Overpassing this barrier is not realistically possible by thermal excitation. But even if it was, we would not escape one of the aspects of microscopic reversibility, that the 1 → A and the 3 → A (via 2) processes have the same rate (the rates depend only on the elevation difference, and not on the slope of the curve). To make the molecule work, one has to rely to a photochemical excitation, which brings enough energy and does not obey microscopic reversibility [2]. Thus excitation of 1 brings the system to an excited state, followed by relaxation towards the avoided crossing region (AC) and de-excitation to the top of the barrier (point A) with the possibility to come back to the initial state 1 or progress towards 2. If 2 has been reached, a thermal excitation transforms 2 into 3, a process which is irreversible as anticipated from the energy difference between 2 and 3 and the low barrier. At this stage, a half-turn has been achieved, and an additional sequence of excitations is necessary to complete a 360° unidirectional turn.

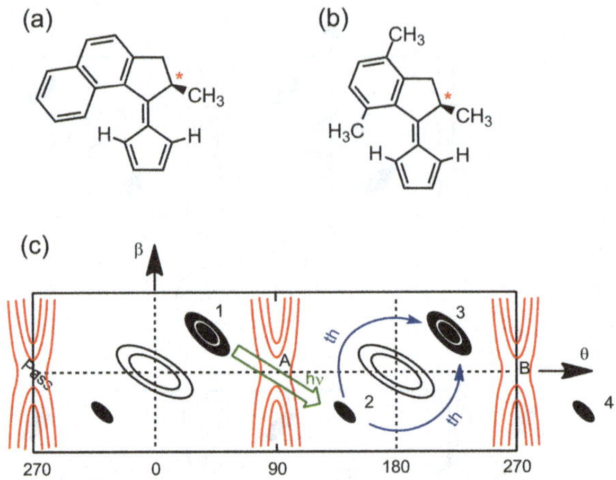

Fig. 11 Minimal Feringa-type molecules: an overcrowded alkene incorporating a chirality center. **a** Pentadiene bottom with a fused naphthalene top group, conferring helicene properties **b** Pentadiene bottom with a fused p-xylene top group **c** Simplified Potential Energy Surface. In black: low energy basins. Green: photochemical step allowing evolution from the lowest basin 1 to a higher elevation 2. Blue: thermal evolution between basin 2 and basin 3 (equivalent to 1) completing a half-turn (two possible trajectories)

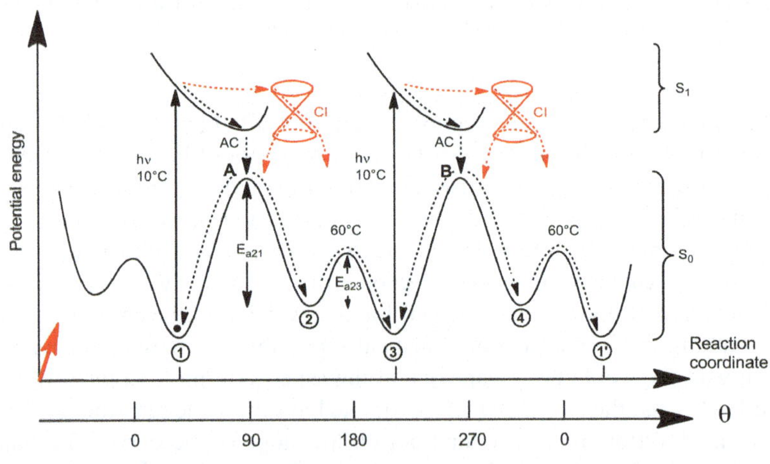

Fig. 12 Energy as function of a reaction coordinate (essentially twisting around the central double bond) for a Feringa-type molecule. Another abscissa with the θ twist angle is also represented, but is not linearly related to the reaction coordinate. AC: Avoided Crossing. In red: Conical Intersections (CI) skewing the photochemical evolution in a definite direction. See text for discussion

As such, the process presents however a weak point, the photochemical conversion. It is well known that, since the system has a choice once arrived on the top of the barrier, the quantum yield for photoisomerization of the double bond is generally close to 0.5. This is due to the symmetry of the potential energy curves which derives, as a first approximation, from the resolution of the Hamiltonian of a two-state system. But in the present case, quantum chemical calculations show a value near 0.9 in the forward direction, which is attributed to the intervention of conical intersections between the ground and first excited state [24, 25]. Let us recall that conical intersections can act as funnels and orient the de-excitation in a definite direction. In the present case, conical intersections (several of them are identified by calculations) can be imagined as residing in an additional dimension, perpendicular to the plane of the Fig. 12, and, due to chirality, their position can be "skewed" with respect to the AC region. This fortunate circumstance, combined with the irreversibility of the thermal step, is a key factor for the success of unidirectional motion.

5 Flashing Ratchet Mechanisms and Symmetry

5.1 Flashing Ratchet Concept

A particularly efficient scheme for the generation of a unidirectional motion is the "flashing ratchet" involving the periodic switching of a potential energy profile between an "ON" state with asymmetric teeth and an "OFF" flat state (Fig. 13) [26, 27].

This mechanism is pertinent for the evolution of a population of particles subject to Brownian motion at a micron scale, while we are concerned here with motions of unique molecules at the nanometer scale. However, the general mathematical treatment can present similarities. In our case, the curves on Fig. 13 could represent the potential energy of a molecule as a function of a geometrical parameter (internal torsion angle, position on a surface, and so on). At the scale of a unique molecule,

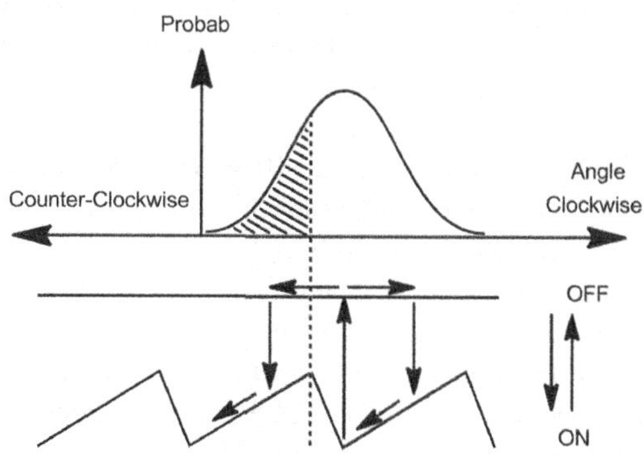

Fig. 13 The simple flashing ratchet mechanism. Below: Two potential energy curves between which he system evolves, a lower asymmetric one, and an upper flat one. Above: Probability for reaching a given angle, taking into account the spreading of the probability function with time. See text for discussion

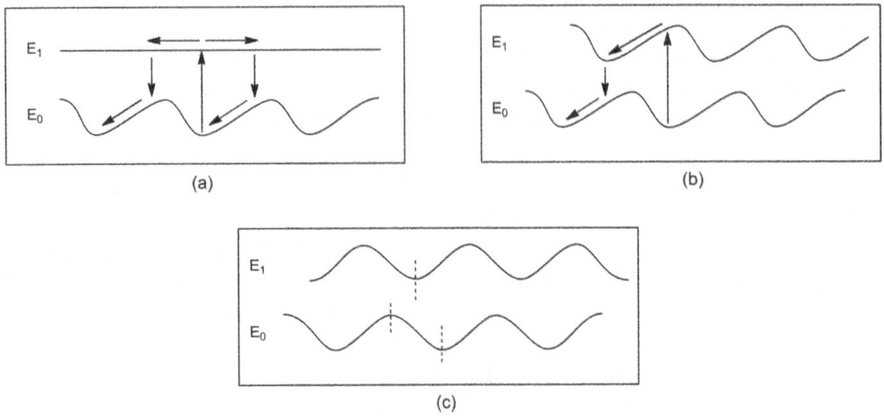

Fig. 14 Some dispositions of potential energy curves in flashing ratchets mechanisms. **a** Simple case corresponding to Fig. 13. **b** Two asymmetric sawtooth profiles shifted in such a way that a left-wise motion is obtained by two consecutive downhill drifts without need for a diffusion step, a more efficient disposition than in (**a**) [27]. **c** Two shifted *symmetric* profiles but an asymmetry (not shown) in the transition rates between E_0 and E_1. See text

energy spreads with time on all degrees of freedom, and so does the probability for a parameter to reach a given value.

Starting from the "ON" situation (Fig. 13, bottom), if one switches to the "OFF" flat state, the probability curve for presenting a given angle, initially narrow, spreads with time by a diffusion-like process. When returning to the "ON" state, there is trapping in one of the potential wells, but due to the asymmetry of the teeth, there are more chances for the system to have evolved towards the left. After several cycles, there is a net motion towards the left, incidentally counter to the intuitive motion of a macroscopic device. There is an optimum frequency for the switching. If it is too high, the system has no time to move; if it is too low, the probability function spreads on too many teeth on both sides. Thus the time structure of fluctuations plays a role (in particular it must not be generated by a white noise). To summarize, according to Magnasco [28], "Broken symmetry and time correlations are sufficient ingredients for transport". For an extensive discussion of the behavior of Brownian ratchets, the role of symmetries and Curie principle, the importance of the time structure of the excitation, the extension to quantum aspects, see [26].

Many variants of this mechanism showing how a random process can lead to unidirectional motion are available (see Fig. 14 for a few examples). In particular one can switch between different sawtooth profiles with different asymmetries and/or minima/maxima positions [27]. It is even possible to achieve unidirectional motion with two *symmetrical* energy profiles (a ground state and an excited state) but suitably shifted with respect to each other [29]. In this last case the achievement of unidirectional motion is not apparent in Fig. 14c, because it results from different transition rates between E_0 and E_1. The detailed theory justifying this surprising result is based on Fokker-Planck equation [29].

5.2 *Ratchets and Potential Energy Curves or Surfaces in STM Studies*

Three examples below show the role of asymmetry (chirality) in the occurrence of unidirectional rotation of molecules deposited on surfaces and observed and excited by STM. We can use the semi- classical concepts developed above because, although molecules are quantum objects, once deposited on the surface and subject to a flux of tunneling electrons, they recover partially a classical behavior.

The STM is an efficient tool to bring energy in a very local way, essentially by two mechanisms: electric field effect, and inelastic electron tunneling [22, 30]. In the first case, the tip position is such that a high bias voltage can be imposed with a low tunnel current (pA range). The resulting electric field interacts with the dipole moment (permanent or not) of the molecule. Conversely, for inelastic tunneling, the bias voltage is smaller and the tunneling current is large (nA range). There are thus a large number of electrons crossing the molecule per unit time. In most cases, they go through the adsorbate without releasing energy (elastic process), but a small proportion, about 1/10 000 can lead to an inelastic scattering. Some energy is deposited in the molecule and can achieve a molecular motion or displacement.

The Platinum-porphyrin molecule Pt-MTBPP (5-(4-methoxy carbonylphenyl)-10, 15, 20-tris(3, 5-di-*tert*-butylphenyl) porphinatoplatinum(II)) can be deposited on a Au(111) surface and forms chiral dimers with a specific shape like an S or Ƨ (Fig. 15) [31]. One of the two porphyrins is pinned and the other only loosely linked to the surface, allowing a rotational motion of the dimer. (Note that the free dimer would present a plane of symmetry and thus no chirality, this latter property is in fact a consequence of bonding to a surface-something like planar chirality, here called "surface chirality").

When applying a +2 V pulse by the tip of a STM located on the unpinned part of the dimer, unidirectional rotation around the pinning center by up to 60° is observed in the clockwise direction for the S isomer and in anti-clockwise direction for the Ƨ one. Thus the surface chirality determines the rotation direction. The proposed mechanism is a shot-noise inelastic effect [22, 30]. It involves the potential energy curves of the ground state and of an excited state, the latter being shifted with respect to the former by the surface chirality of the complex. Interestingly, it is possible to reverse the chirality of the dimer by applying a negative voltage pulse (−1.9 V), although the probability of such events is relatively modest (8%).

As a second example, the "piano stool" molecule represented on Fig. 16 is made of two parts able to rotate with respect to one another thanks to the presence of a ruthenium atom acting as a joint [32]. One of the parts is made of the three indazole groups (the legs of the piano stool), the other of a pentabromophenyl–cyclopentadienyl (the seat). On a Au(111) surface, the molecule adsorbs "upside down", i.e. with the three legs upward (despite the presence of thioether groups which should favor linking the legs to the surface). Due to steric crowding between adjacent phenyl rings, all the phenyl rings are tilted in the same direction, thus imparting a chirality of the atropoisomerism type (see Fig. 3). This chirality is in turn transferred to the set of

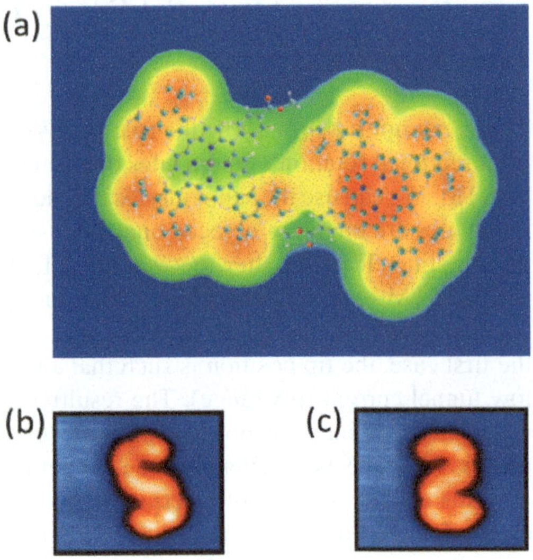

Fig. 15 Pt-porhyrin dimer on Au(111) surface. **a** Calculated STM image of a dimer with the right porphyrin pinned by an adatom. **b** and **c** Enantiomeric forms of the dimers with S or Ƨ shapes respectively. Adapted with permission from Mishra et al., *Nano Lett.* 2015, 15, 4793 [31]. Copyright 2015 American Chemical Society

three indazole groups, and the final result is a typical shape in STM images exhibiting C_3 symmetry (Fig. 16c). The relative positions of the two parts have been analyzed elsewhere [33]. Due to the 3-fold symmetry of the legs and the 5-fold symmetry of the seat, the energy takes the same value after 1/15 of a turn (a change in 24° of the torsion dihedral angle).

On Au(111), only two enantiomers are observed, with equal probabilities, denoted M and P. Upon application of suitable voltage pulses, unidirectional rotation is observed, the M and P forms rotating preferentially clockwise and anti-clockwise respectively. The rotation angle is 1 or 2 elementary step of 24°. The two excitation modes mentioned above, electric field effect and inelastic tunneling were found efficient.

The third example involves a molecule of the same family as the piano stool above, but the Bromine atoms of the phenyl arms have been replaced by Ferrocene groups. Moreover, in order to detect rotation, four arms only are functionalized, the remaining (truncated) one being use as a "marker" [34]. On Au(111), the piano stool now stands "upright", linked to the surface by the thioether groups of the three legs (Fig. 17).

Upon excitation with a STM tip, unidirectional rotation has been observed [34]. Moreover, the rotation direction could be reversed according to the exact tip position. With the tip above the truncated arm, the rotation was clockwise, while with the tip above a ferrocene arm, it was counterclockwise. Assuming excitation by inelastic tunneling, theoretical studies justify this behavior by the different properties of the ground and excited state and the different molecular orbitals involved, yielding displaced potential energy curves (Fig. 17e, d). Incidentally, in addition to the modulation every 1/15 of a turn as in the previous case, one can note a general modulation

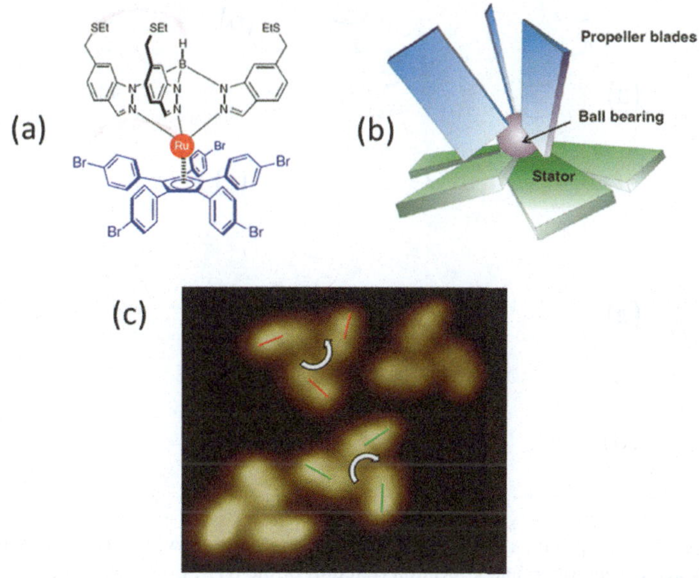

Fig. 16 The Piano stool molecule upside down. **a** Chemical structure. **b** Simplified scheme showing the "seat" part with 5-fold symmetry and the "legs" with 3-fold symmetry. **c** STM images of the legs (propeller blades) and their unidirectional motion determined by chirality: upper molecule counterclockwise, lower molecule clockwise. Reproduced with permission of Springer Nature from Zhang et al., *Nature Comm.* 2019 [32] under Creative Commons License

of about 120° (1/3 of a turn) corresponding to the passage of the truncated arm in front of one of the three legs.

It is however puzzling that a given directionality could be obtained with such a structure which does not present at first sights a center of asymmetry. But the previous example shows the possible occurrence of two atropoisomers due to the correlated orientation of the phenyl rings, and in the present case, there can be also different orientations of the ferrocene groups. It is thus possible that the clockwise/anti-clockwise rule formulated above can be reverted from one molecule to the other.

6 Concluding Remarks About Symmetry

In this chapter we started from simple symmetry considerations about the chirality of the systems, in the spirit of Curie's Principle. A nice illustration was provided by the rattle-back toy. We then moved to potential energy curves and their symmetry (or asymmetry) when energy is plotted as a function of a rotation parameter (potential energy profile). This shape is of course governed by the geometrical structure of the system, in particular its chiral or achiral character. The flashing ratchet paradigm involving switching or more generally evolution between different potential energy

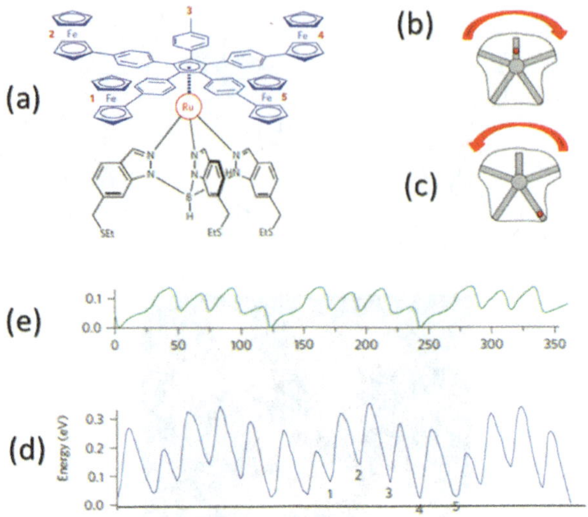

Fig. 17 A piano stool upright. **a** Chemical structure showing the seat with 5 arms, one of them being truncated to detect rotation. **b** and **c** location (red dot) of the STM tip for excitation, either on the truncated arm or on a normal arm, with the corresponding rotation directions. **d** and **e** Calculated potential energy curves as functions of rotation angle for the ground state and the excited state respectively. Reproduced with permission of Nature from Perera et al., *Nature Nanotech.* 2013 [34]; permission conveyed through Copyright Clearance Center, Inc

curves is of great use to explain at least qualitatively the behavior of many systems, including the case of molecules addressed by a tunnel current. In addition, as mentioned in Sect. 4.5.2 of [26], Feringa-type molecules can also be considered as examples of pulsating ratchets. We can note that the evolution between states 1 and 3 on Fig. 12 has a general asymmetric shape. Of course, we cannot use directly the flashing ratchet scheme in this case because the abscissa (a reaction coordinate) is ill-defined, but broadly speaking, the existence of the intermediate state 2 breaks the symmetry around the highest point, A or B, of the potential energy curve.

The energy profile is thus extremely important and can be addressed by theoretical calculations showing a strong asymmetry in some cases [22, 35]. However in other cases, the profile is extremely close to symmetrical, even when the chemical structure presents asymmetry or chirality. This is in particular the case with DFT calculations using the Nudged Elastic Band method [36]. This apparent contradiction could come from the use of a reaction coordinate as abscissa, which is a complex combination of several internal coordinates, with eventually their relative weights varying along a given profile. In fact, the modelization of the potential energy of complex molecules requires several nuclear coordinates, so that we are concerned with Potential Energy Surfaces or Hypersurfaces. In other words, in usual energy profile calculations, we project the complex behavior on a simpler coordinate frame in an unknown and uncontrolled way. But in the case of only two nuclear coordinates, it was shown several decades ago, that for a simple reaction involving a combination of rotation and inversion of the molecular skeleton, the energy profile could be symmetric or asymmetric according to the chosen trajectory in nuclear configuration space [37]. Ratchet models in more than one dimension would certainly be worth developed.

Acknowledgements Helpful discussions with Prof. Klaus Hornberger and Dr. Christian Joachim are gratefully acknowledged.

References

1. Curie, P.: Sur la symétrie dans les phénomènes physiques, symétrie d'un champ électrique et d'un champ magnétique. J. Phys. Théor. et Appl. **3**, 393–415 (1894). (For a recent discussion, see: Ismael, J.: Curie's Principle. Synthese **110**, 167–190 (1997))
2. Astumian, R.D.: Microscopic reversibility as the organizing principle of molecular machines. Nat. Nanotechnol. **7**, 685–688 (2012)
3. Case, W., Jalal, S.: The rattleback revisited. Am. J. Phys. **82**, 654–658 (2014). (A video of its behavior can be found on http://www.youtube.com/watch?v=69Xm762qE8o)
4. Eliel, E.L., Wien, S.H.: Stereochemistry of Organic Compounds. Wiley (1994). ((a) pp 19, 104; (b) p 1142)
5. Klein-Hessling, C., Sünkel, K.: (η^4-Tetraphenylcyclobutadiene)(η^5-pentaphenylcyclopentadienyl)-cobalt. Molbank **2022**, M1502 (2022). and references therein
6. Barron, L.D.: True and false chirality and absolute asymmetric synthesis. J. Am. Chem. Soc. **108**, 5539–5542 (1986). Owens, A., Yachmenev, A., Yurchenko, S.N., Küpper, J.: Climbing the rotational ladder to chirality. Phys. Rev. Lett. **121**, 193201 (2018)
7. Hund, F.: Z. Phys. **43**, 805 (1927)
8. Quack, M., Seyfang, G., Wichmann, G.: Perspectives on parity violation in chiral molecules: theory, spectroscopic experiment and biomolecular homochirality. Chem. Sci. **13**, 10598–10643 (2022)
9. Bersuker, I.B.: Jahn-Teller and Pseudo Jahn-Teller Effects: from particular features to general tools in exploring molecular and solid state properties. Chem. Rev. **121**, 1463–1512 (2021)
10. Gorinchoy, N.N., Ogurtsov, I.Y., Arsene, I.: Vibronic origin of the "skewed" anticline configuration of the hydrogen peroxide molecule. Chem. J. Moldava. Gene. Indus. Ecolog. Chem. **3**, 105–111 (2008)
11. Maciel, G.S., Barreto, P.R.P., Palazzetti, F., Lombardi, A., Aquilanti, V.: A quantum chemical study of H_2S_2: intramolecular torsional mode and intermolecular interactions with rare gases. J. Chem. Phys. **129**, 164302 (2008)
12. Sahu, N., Richardson, J.O., Berger, R.: Instanton calculations of tunneling splittings in chiral molecules. J. Comput. Chem. **42**, 210–221 (2021)
13. Joos, E.: Elements of environmental decoherence. In: Blanchard, P., Guilini, D., Joos, E., Kiefer, C., Stamatescu, I.-O. (eds.) Decoherence: Theoretical, Experimental, and Conceptual Problems. Proceedings of Bielefeld Conference. Springer (1999)
14. Trost, J., Hornberger, K.: Hund's paradox and the collisional stabilization of chiral molecules. Phys. Rev. Lett. **103**, 023202 (2009)
15. Laidig, K.E.: Atomic origins of molecular polarizabilities. Can. J. Chem. **74**, 1131–1138 (1996)
16. Trost, J., Hornberger, K.: The twisted arc model for chiral molecules. Chem. Phys. **335**, 115 (2007)
17. See the pedagogical video by Sabine Hossenfelder. https://www.youtube.com/watch?v=igsuIuI_HAQ
18. Bahrami, M., Bassi, A.: On the tunneling properties of non-planar molecules in a gas medium. Phys. Rev. A **84**, 062115 (2011)

19. Zurek, W.H.: Quantum Darwinism. Nat. Phys. **5**, 181–188 (2009)
20. Zurek, W.H.: Preferred states, predictability, classicality and the environment-induced decoherence. Progr. Theoret. Phys. **89**, 281–312 (1993)
21. Grein, F.: Twist angles and rotational energy barriers in biphenyl and substituted biphenyls. J. Phys. Chem. A **106**, 3823–3827 (2002)
22. Echeverria, J., Monturet, S., Joachim, C.: One-way rotation of a molecule-rotor driven by a shot-noise. Nanoscale **6**, 2793–2799 (2014)
23. Ammon, H.L., Wheeler, G.L., Agranat, I.: The structure of Octachloro-pentafulvalene. Tetrahedron **29**, 2695–2698 (1973)
24. Pollard, M.M., Meetsma, A., Feringa, B.L.: A redesign of light-driven rotary molecular motors. Org. Biomol. Chem. **6**, 507–512 (2008). Kazaryan, A., Kistemaker, J.C.M., Schäfer, L.V., Browne, W. R., Feringa, B.L., Filatov, M.: Understanding the dynamics behind the photoisomerization of a light-driven fluorene molecular rotary motor. J. Phys. Chem. A **114**, 5058–5067 (2010)
25. Li, Y., Liu, F., Wang, B., Su, Q., Wang, W., Morokuma, K.: Different conical intersections control nonadiabatic photochemistry of fluorine light-driven molecular rotary motor: a CASSCF and spin-flip DFT study. J. Chem. Phys. **145**, 244311 (2016)
26. Reimann, P.: Brownian motors: noisy transport far from equilibrium. Phys. Rep. **361**, 57–265 (2002)
27. Jülicher, F., Ajdari, A., Prost, J.: Modeling molecular motors. Rev. Mod. Phys. **69**, 1269–1281 (1997)
28. Magnasco, M.O.: Forced thermal ratchets. Phys. Rev. Lett. **71**, 1477–1481 (1993)
29. Sanchez-Palencia, L.: Directed transport of Brownian particles in a double symmetric potential. Phys. Rev. E **70**, 011102 (2004)
30. Jacquot de Rouville, H.-P., Adrouche, S., Bouju, X., Launay, J.-P., Rapenne, G., Joachim, C.: From early prototypes to on-surface drivable single molecule Nano-vehicles. In: Moresco, F., Joachim, C. (eds.) Single Molecule Mechanics on a Surface. Advances in Atoms and Single Molecular Machines, pp. 59–82. Springer (2023)
31. Mishra, P., Hill, J.P., Vijayaraghavan, S., Van Rossom, W., Yoshizawa, S., Grisolia, M., Echeverria, J., Ono, T., Ariga, K., Nakayama, T., Joachim, C., Uchihashu, T.: Current-driven supramolecular motor with in situ surface chiral directional switching. Nano Lett. **15**, 4793–4798 (2015)
32. Zhang, Y., Calupitan, J.P., Rojas, T., Tumbleson, R., Erbland, G., Kammerer, C., Tolulope, M.A., Wang, S., Curtiss, L.A., Ngo, A.T., Ulloa, S.E., Rapenne, G., Hla, S.W.: A chiral molecular propeller designed for unidirectional rotation on a surface. Nat. Commun. **10**, 3742 (2019)
33. Carella, A., Launay, J.-P., Poteau, R., Rapenne, G.: Synthesis and reactivity of [Penta(4-halogenophenyl)cyclopentadienyl]-[hydrotris(indazolyl)borato] ruthenium(II) complexes; rotation-induced Fosbury flop in an organometallic molecular turnstile. Chem. Eur. J. **14**, 8147–8156 (2008)
34. Perera, U.G.E., Ample, F., Kersell, H., Zhang, Y., Vives, G., Echeverria, J., Grisolia, M., Rapenne, G., Joachim, C., Hla, S.-W.: Controlled clockwise and anticlockwise rotational switching of a molecular rotor. Nat. Nanotechnol. **8**, 46–51 (2013)
35. Kelly, T.R., Tellitu, I., Sestelo, J.P.: In search of molecular ratchets. Angew. Chem. Int. Ed. **36**, 1866–1868 (1997)
36. Robles, R., Lorente, N.: Asymmetric energy barriers in unidirectional molecule-rotors. In: Bouju, X., Joachim, C. (eds.), Crossroad of Maxwell Demon. Advances in Atom and Single Molecule Machines (Chap. 8, in this volume), pp. 159–177
37. Wolfe, S., Schlegel, H.B., Csizmadia, I.G., Bernardi, F.: The chemical dynamics of symmetric and asymmetric reaction coordinates. J. Am. Chem. Soc. **97**, 2020–2024 (1975)

Asymmetric Energy Barriers in Unidirectional Molecule-Rotors

Roberto Robles and Nicolás Lorente

Abstract By applying the nudged elastic band method in the framework of ab initio density functional theory we calculate the energy barriers of unidirectional molecule-rotors. For the rotation to be unidirectional, these barriers must be asymmetric. We study two systems. The first is an acetylene molecule which rotates on a basin formed by a Pd trimer on a chiral PdGa surface. The second are DMBI derivatives on the (111) surface of gold, which rotate after the formation of a strong chemical bond which provides an axle for the rotation. We show that the applied theoretical method is able to produce asymmetric barriers for both systems, although the barriers are too high compared to the experimental values. In addition, different details of the systems are discussed in connection with the experimental results.

1 Introduction

Among the numerous examples of molecular machines one particular class consists in molecules that rotate in a given direction [1–14]. These unidirectional molecule-rotors require two main ingredients: thermodynamic nonequilibrium and an asymmetric rotation potential. The first ingredient is due to the microscopic reversibility principle [15], which forbids unidirectional movements of any system in thermodynamic equilibrium. The nonequilibrium can be achieved by external energy sources like light [2], chemical reactions [1] or the inelastic electrons coming from the tip of a scanning tunneling microscope (STM) [3, 12]. The second ingredient, an asymmetric rotation potential, is in general produced by the absence of inversion symmetry in the

R. Robles (✉) · N. Lorente
Centro de Física de Materiales CFM/MPC (CSIC-UPV/EHU), Paseo Manuel de Lardizabal 5, 20018 Donostia-San Sebastián, Spain
e-mail: roberto.robles@csic.es

N. Lorente
e-mail: nicolas.lorente@csic.es

N. Lorente
Donostia International Physics Center (DIPC), Paseo Manuel de Lardizabal 4, 20018 Donostia-San Sebastián, Spain

X. Bouju and C. Joachim (eds.), *Crossroad of Maxwell Demon*, Advances in Atom and Single Molecule Machines, https://doi.org/10.1007/978-3-031-57904-2_8

system. In the case of molecules adsorbed on surfaces, the lack of inversion symmetry can be achieved by chiral molecules on achiral surfaces [8], by achiral molecules on chiral surfaces [11] or, more generally, by achiral molecules which become chiral after its adsorption on achiral surfaces [12, 16] (the so-called prochiral molecules).

Here we will study the asymmetric rotational energy barriers in the potential energy surface (PES) which are at the origin of unidirectional rotations in molecule-rotors. In particular, we will use the nudged elastic band method within density functional theory. Our goal is to assess the ability of this ab-initio method to calculate such energy barriers.

The chapter is organized as follows. After introducing the theoretical method in Sect. 2, in Sect. 3 we will show that it is able to produce perfectly symmetric energy barriers for achiral molecules in the gas phase; in Sect. 4 we will study the energy barrier of a simple molecule in a chiral surface; and in Sect. 5 we will explore the behavior of DMBI derivatives supported on a Au(111) surface.

2 Theoretical Method

Calculations in this chapter have been done using density functional theory (DFT) as implemented in the VASP code [17]. The projector augmented-wave method has been used to treat the core electrons [18]. We have used a plane-wave basis set with an energy cutoff of 500 eV (400 eV for DMBI derivatives in Sect. 5). For the exchange and correlation functional we have used the PBE flavor of the generalized gradient approximation [19]. In Sect. 3 we have performed additional calculations using the hybrid B3LYP functional [20]. The PBE functional fails to correctly describe long-range interactions, so we have included van der Waals (vdW) interactions following the methods of Tkatchenko-Scheffler [21] and Ruiz et al. [22] for C_2H_2 on PdGa (Sect. 4), and the DFT-D3 method [23] with Becke-Johnson damping [24] for DMBI derivatives (Sect. 5).

Surfaces have been represented by finite slabs. For the PdGa:A$(\bar{1}\bar{1}\bar{1})$Pd$_3$ surface in Sect. 4 we have used a 2×2 unit cell with 12 layers in the $(\bar{1}\bar{1}\bar{1})$ direction. The six bottom layers were kept frozen at their bulk positions during the geometrical relaxations. A vacuum layer of 20 Å was added to avoid spurious interactions between repetitions of the slabs. The Au(111) surfaces in Sect. 5 have been modeled using a $8\times5\sqrt{3}$ rectangular unit cell with 4 layers in the (111) direction, and a vacuum layer of 18 Å. The two bottom layers were kept frozen. Geometries have been optimized until all the forces were smaller than 0.01 eV·Å^{-1}.

Energy barriers have been computed using the nudged elastic band (NEB) method [25]. NEB is an efficient method to find minimum energy paths between two different positions in the potential energy surface (PES). The method optimizes a number of intermediate images along the reaction path until they reach the minimum possible energy. A constraint is introduced consisting in a spring force tangential to the trajectory which ensures that the images remain equidistant. We have used the improved tangent definition of Henkelman et al. [26], and the climbing image modification [27], which forces the highest energy image to be at the saddle point.

We have calculated charge transfers between the adsorbed molecules and surfaces by performing Bader analyses [28]. We have simulated STM images using the Tersoff and Hamann theory [29] following the method by Bocquet et al. [30] as implemented in STMpw [31]. VESTA [32] and VMD [33] have been used to generate structural and density images.

3 Rotation of 2,2'-Dibromobiphenyl in the Gas-Phase

The biphenyl (BP) molecule is formed by two phenyl rings joint by a C-C single bond. Biphenyl is twisted, with a dihedral angle between phenyl rings of $44.4 \pm 1.2°$ according to electron diffraction measurements [34]. This value of the angle is the result of the competition between the effect of π-electron delocalization, which would favor the coplanar geometry, and the steric repulsion of the hydrogens at the ortho positions, which would favor a 90° dihedral angle. Substitution of the ortho hydrogens by larger atoms leads to the enhancement of the steric effect and, hence, to bigger dihedral angles. In the case of 2,2'-dibromobiphenyl (DBBP) two ortho hydrogens are substituted by Br atoms, leading to an angle of $84.8 \pm 2°$, according to X-ray data [35].

The rotation around the central single bond of BP and its derivatives is a prototypical case for the study of internal rotation in molecules. Here we will study DBBP, which due to the presence of two Br atoms shows a significant barrier height of more than 1 eV [36]. This high value is due to the repulsion of the two Br atoms and presents an important challenge to the calculation of the barrier shape using NEB.

The first step in a NEB calculation is always a careful determination of the two local minima which will be the extrema of the NEB path. By symmetry, in this case there are two identical minima at ϕ and $-\phi$ dihedral angles. We calculated the dihedral angle between atoms C_2, C_1, $C_{1'}$, $C_{2'}$ as shown in the inset of Fig. 1. Using the PBE functional we found an optimized dihedral angle of 87.40°. This value compares well with the experimental angle of $84.8 \pm 2°$ [35] and a value of 91.54° calculated using B3LYP/6-311+G* [36]. As stated before, by symmetry a second minimum can be found at $-87.40°$. Typically the NEB method is used to find transition states (TSs), but in this case the symmetry of the molecules allow us to know a priori that there are at least two TSs between the two local minima located at 0° and 180° dihedral angles. We found the first state 1.395 eV higher in energy than the ground state (GS), while the second one is at 0.720 eV. The difference between both TSs can be explained by the steric repulsion being higher for Br-Br than for Br-H. Previous B3LYP/6-311+G* calculations found both minima at 1.585 eV and 0.866 eV, respectively [36]. The difference can be explained by the use of different functionals. Actually we have repeated our calculations with B3LYP, finding a much closer value of 1.523 eV for the 0° case.

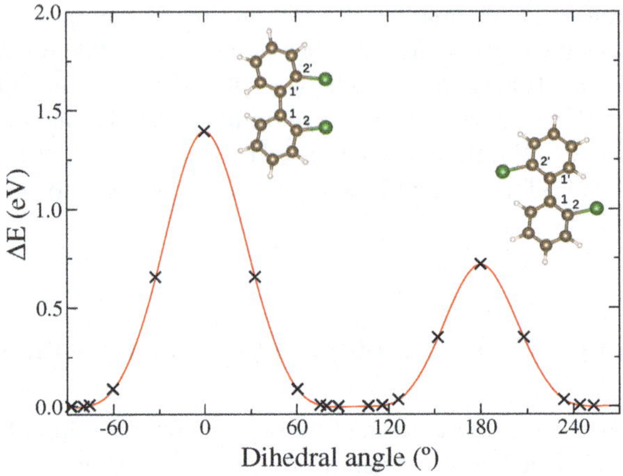

Fig. 1 Energy barrier for the rotation of DBBP between $-90°$ and $270°$. It has been computed by two different NEB calculations with 9 images each using the PBE functional. The line is a cubic spline fitted to the data. In the insets we show ball and stick models for the TS at 0° (left) and 180° (right) (C, Br and H atoms are represented in brown, green and pink). The atoms used to calculate the dihedral angle $\phi(C_2C_1C_{1'}C_{2'})$ are shown in the insets

In the insets of Fig. 1 we can see that there is a big geometrical distorsion at 0°, where the phenyl rings, while in the same plane, are clearly rotated with respect to each other. It is also worth to note that in this case the NEB method struggles to accurately find the transition state if no symmetry considerations are taken into account. Doing that calculation we obtained the TS at 14° instead of 0°, with an energy of 1.318 eV, 77 meV lower than the true value.

Once the starting and final points of the NEB calculation have been established we have set up the calculation with 9 intermediate images and taking into account a coplanar geometry for the TSs. We have performed two separate calculations for a clockwise rotation between ϕ and $-\phi$ for the 0° TS, and between $-\phi$ and ϕ for the 180° one. Both calculations are jointly shown in Fig. 1 for the PBE functional. Two perfectly symmetric barriers are found around each TS. This is the expected behavior given the symmetry of the molecule. It is interesting to note that the points along the NEB trajectory are clearly not equidistant, as they should be after a NEB calculation. The reason is that we have represented the barrier as a function of the dihedral angle, while in the calculation the distance between images is computed using the square root of the squared difference of the coordinates of the images. Using this distance to represent the reaction coordinate the images are indeed equidistant.

We have repeated the calculation for the B3LYP functional around the 0° TS. In this case the optimized dihedral angle of the minima is 88.74° and the barrier height is 1.523 eV, as indicated previously. The barrier can be seen in Fig. 2 and it shows a similar shape to the barrier calculated using the PBE functional (Fig. 1).

In order to rotate, the molecule must receive some energy to overcome the energy barrier. This can be done for example by the injection of electrons or holes by a

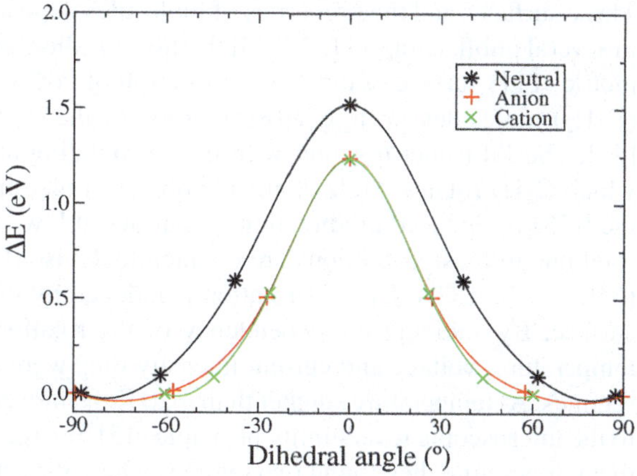

Fig. 2 Energy barriers for the rotation of DBBP between −90° and 90° for the neutral (black stars), anion (red crosses) and cation (green exes). They have been computed using 5 images with the B3LYP functional. The lines are cubic splines fitted to the data

STM tip. In this process the molecule might have a transient state in which it is charged forming an anion or a cation. The existence of this transient state might change the barrier. In order to check this possibility we have calculated the barrier for the anion and the cation with the B3LYP functional. For the anion the optimized dihedral angle is 89.41°, quite similar to the neutral case. However, for the cation the angle is quite different at 60.38°. A lower value might indicate a stronger π-electron delocalization. The barrier height is decreased for both charged molecules around a 20% to 1.239 eV for the anion and 1.237 eV for the cation. The barriers are displayed in Fig. 2 and show a similar shape for the neutral and the charged molecules. This result exemplifies that the existence of charged transient states might be important in the description of the barriers and might explain differences between theory and experiment.

The results for DBBP prove that the NEB method allows us to obtain energy barriers which keep the correct symmetry, but careful calculations must be performed. We have also shown that the height and shape of the barriers would depend on the chosen functional and the charge state of the molecule.

4 Rotation on a Chiral Surface

In this chapter we explore the energy barriers for a system in which the asymmetry is due to the surface itself rather than to the rotating molecule. Following a macroscopic analogy, our system would be a motor where the rotor is symmetric (acetylene), while the stator is chiral (the surface of a noncentrosymmetric crystal). This approach has the advantage of, at least theoretically, work for any molecule, even symmetric ones.

The system was experimentally studied by Stolz et al. [11] using STM. For the chiral surface they chose the intermetallic compound PdGa with 1:1 stoichiometry. PdGa is a noncentrosymmetric crystal whose bulk terminated surfaces are chiral.

These surfaces and the properties of molecules deposited on them have been studied in several publications [11, 37–45]. In this case the chosen surface is a threefold symmetric $(\bar{1}\bar{1}\bar{1})$ surface of the A enantiomorph of PdGa [38]. On this surface acetylene (C_2H_2) molecules strongly adsorb on top of the Pd trimers present on the surface [42]. The Pd trimer together with the surrounding atoms form the chiral stator on which C_2H_2 rotates. Stolz et al. [11] observed that inelastic electrons coming from the STM tip induced unidirectional rotations at low voltages and temperatures. The rotations in these conditions are counterclockwise (CCW) with a directionality up to 97%. The CCW sense of rotation is induced by the underlying symmetry of the surface. By studying the dependency of the rotation on different parameters like temperature, voltage and current intensity they were able to identify three different regimes. At temperatures higher than ≈20 K the movement is thermal and, according to the microscopic reversibility principle [15], the rotation is non-directional. Below that temperature they found two regimes where the rotation is unidirectional. First, a classical regime (CR) where the frequency of the rotation depends on the parameters above. Then, for temperatures below 15 K, voltages smaller than 30 mV and currents under 200 pA, they found a third regime (TR), where the frequency is independent of the parameters. They could only explain the TR by considering quantum tunneling through the barrier. Analyzing the dependence of the rotation at low voltages as a function of temperature they found an Arrhenius behavior. From the corresponding Arrhenius plot they were able to extract a barrier height of $\Delta E_B = 27.5 \pm 7.1$ meV. Moreover, they proposed a Langevin kinetic model from which they extracted a barrier asymmetry R_{asym} between 1.25 and 1.5. They also determined that the rotation is best described as a "tumbling" rotation, where the center of mass of C_2H_2 moves in a circle with radius 0.5 ± 0.1 Å.

Here we are interested in the asymmetry of barriers, so we will focus in obtaining a barrier for C_2H_2 on the PdGa surface. The details of the rotation on the different regimes, in particular the TR, are well beyond the scope of this chapter.

The first step for the determination of the barrier is an accurate description of the surface. There are several possible surfaces for PdGa depending on the enantiomorph (A or B) and the surface termination. Different possibilities have been studied experimentally [37, 38] and theoretically [39]. Here they grew the $(\bar{1}\bar{1}\bar{1})$ surface of the A enantiomorph, which under their experimental conditions was terminated by a layer containing Pd trimers (PdGa:A$(\bar{1}\bar{1}\bar{1})$$Pd_3$). STM topographic images of the surface typically showed a hexagonal array of protrusions which are formed by the Pd trimers. In the experiment they observed that C_2H_2 is adsorbed on the trimers. The molecule shows three different orientations which are equivalent and rotated by 60°. In Fig. 3 we show the (2×2) unit cell used in the calculation. The cluster which acts as a stator is highlighted in bold colors. The cluster is formed by the Pd trimer on which C_2H_2 rotates, together with its 6 closest Ga neighbors in the next layer and 3 Pd atoms in the layer below. It is easy to see the chirality of the cluster which will govern the directional movement of the molecule.

We have carefully optimized the geometry of two local minima of C_2H_2. The results are shown in the left side of Fig. 4. The final geometries are identical but rotated by 60°. The final energies are also the same within the accuracy of the calculation. In

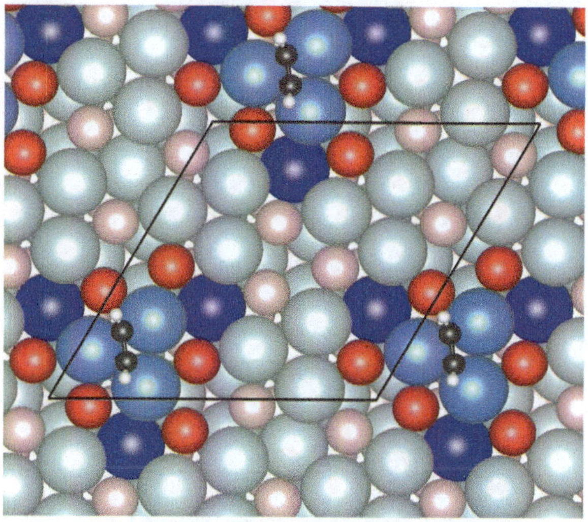

Fig. 3 2×2 unit cell used to simulate C_2H_2 on the $(\bar{1}\bar{1}\bar{1})$ PdGa surface. The cluster which acts as a stator and is responsible of the chirality felt by the molecule is highlighted in bold colors. Blue, red, black and white colors correspond to Pd, Ga, C and H atoms, respectively

order to compare with the experimental results, we have computed the topographic STM image at constant current and at a voltage of 0.2 V. The agreement with the experimental results by Stolz et al. (Fig. 1 C–G of Ref. [11]) is quite good. In the images we can see the molecule formed by two elongated lobes surrounded by an hexagon formed by six protrusions. Comparing left and right sides of Fig. 4 we can assign the hexagon to the Pd trimers surrounding C_2H_2, as indicated by circles in the figure. Two asymmetric dark spots at both sides of the molecule can also be observed in both theory and experiment. In the upper left corner of Fig. 4 we show a lateral image of the relaxed geometry of C_2H_2. The hydrogens are pulled away from the surface, and the molecule losses its gas-phase linear symmetry. The protruding hydrogens are most likely responsible for the two lobes forming the image of the molecule.

Using the calculated local minima of C_2H_2 as starting and final points we have performed a NEB calculation with 7 images to compute the energy barrier. The results are shown in Fig. 5. The first result is that the barrier is clearly asymmetric. In order to emphasize the asymmetry we have plotted in blue the barriers for the CCW (solid blue line) and for the CW (dotted blue line) rotations. In other words, we have plotted the computed trajectory from the initial to the final points and vice versa. We can observe (i) the transition state is not at the center of the trajectory and (ii) the slope of the barrier is not the same at left and right sides of the TS. So the barrier corresponds to a ratchet potential as expected by the symmetry of the system. This result confirms that, qualitatively, a NEB calculation is able to capture the asymmetry of the barrier. From the geometries of the intermediate states we can track the movement of the center of mass of the molecule along the trajectory of the rotation. This result can be seen in the inset of Fig. 5. The center of mass moves along the trajectory indicating that the molecule is indeed a "tumbling" rotor, as discussed

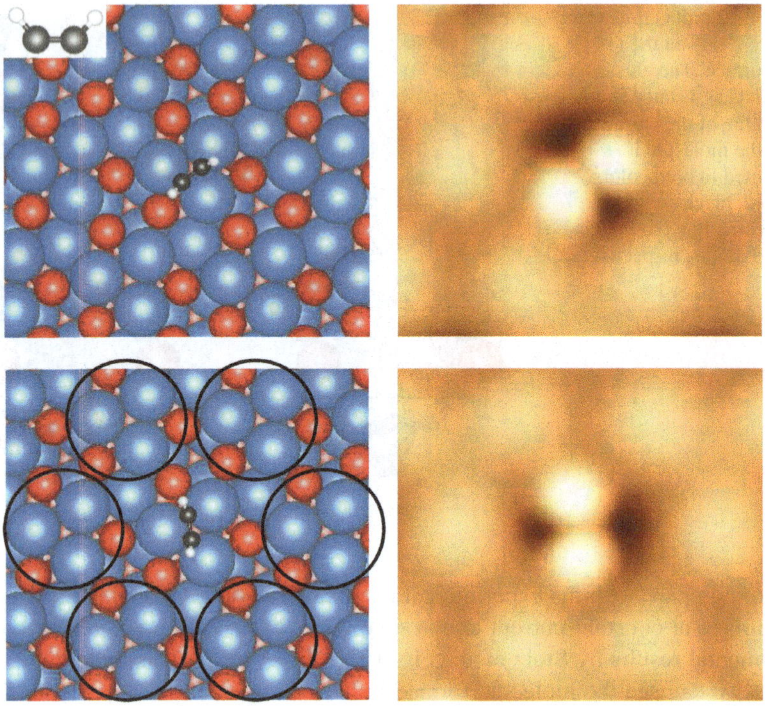

Fig. 4 (Left) Two stable positions of C_2H_2 on the surface used as initial and final points of the NEB calculation. (Right) STM topographic images simulated using the STMpw code [31]. They correspond to the respective geometry on the left. The six protrusions around the molecule correspond to the six Pd trimers around C_2H_2 marked by circles on top of the geometry. In the inset at the upper left corner we show a lateral image of the molecule, where the loss of the gas-phase linear symmetry can be observed

in the experimental paper. However, while they assumed a circular movement of the center of mass, we observe a more complicated shape resembling a Reuleaux triangle with round corners.

By analyzing the barrier we can extract a height of $\Delta E_B = 61$ meV and an asymmetry $R_{\text{asym}} = 1.17$. The experimental values were $\Delta E_B = 27.5 \pm 7.1$ meV and $1.25 < R_{\text{asym}} < 1.5$. So the calculation gives a barrier too high and an asymmetry too small.

In a previous study, Yakutovich et al. [45] studied the enantioselective adsorption of molecules on another PdGa surface (PdGa:A(111)Pd_1). Experimentally, Prinz et al. had observed than the adsorption of 9-ethynylphenanthrene on that PdGa surface showed a quite significant enantiomeric excess, reaching almost an enantiopure ensemble at room temperature [44]. By doing DFT simulations, Yakutovich et al. [45] showed that the experimental result could only be explained with an appropriate description of the vdW interaction, in particular by using the vdW^{surf} method [22]. The results discussed here up to now were obtained using the more common

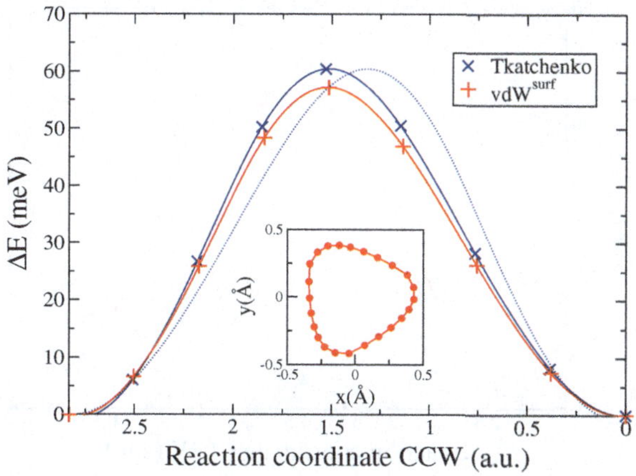

Fig. 5 Energy barrier for the CCW rotation of C_2H_2. The calculation made with the Tkatchenko-Scheffler (vdW^{surf}) vdW approximation is shown in blue (red). The lines correspond to cubic splines fitted to the points. In order to visualize the asymmetry of the barrier we plot with a blue dotted line the barrier for Tkatchenko-Scheffler from the final to the initial points, i.e., the CW rotation. In the inset the movement of the center of mass of the molecule is shown. The line is just a guide to the eye

Tkatchenko-Scheffler method [21] (blue line in Fig. 5), so we repeated the calculation using vdW^{surf}. As can be seen in the red line of Fig. 5, the results with both methods are similar, with a slightly lower barrier ($\Delta E_B = 57$ meV) for vdW^{surf} and a similar asymmetry of $R_{asym} = 1.15$. The smaller difference between both vdW methods in this case compared to reference [45] is probably due to the fact than C_2H_2 is much smaller than 9-ethynylphenanthrene, and therefore the vdW interaction with the surface is much more important for the later.

We have tried to improve the calculation as far as our computational capabilities allow us. In order to do so, we have increased some parameters in the calculation, in particular the k-point mesh from $(3 \times 3 \times 1)$ to $(5 \times 5 \times 1)$, and the number of images in the NEB trajectory, from 7 to 15. The results are shown in Fig. 6. Now the barrier height decreases to $\Delta E_B = 55$ meV, but the asymmetry also does to a value $R_{asym} = 1.06$. To further improve the results we have calculated the contribution of the zero point energy (ZPE) to the barrier. We have computed the ZPE for both GS and TS, resulting in a decreased value for the barrier height of $\Delta E_{B,\mathrm{ZPE}} = 46$ meV. Although improved, the agreement with the experimental value is still poor. A further increase of the accuracy of the calculation would involve approaches beyond DFT which could describe better the PdGa surface and its interaction with C_2H_2, or bigger unit cells to limit the possible remaining interaction between the periodic images of C_2H_2. Any of those options are beyond our current computational resources. Another source of improvement would be the consideration of the effects of the STM tip in the barrier, which has been shown to modify the PES [46].

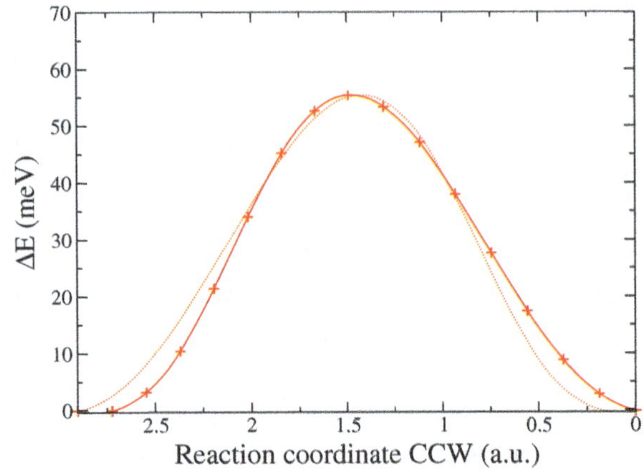

Fig. 6 Energy barrier for the CCW rotation of C_2H_2 (red crosses). The calculation was made with the Tkatchenko-Scheffler vdW approximation, 15 images, and a $(5 \times 5 \times 1)$ k-point mesh. The lines correspond to cubic splines fitted to the points. In order to visualize the asymmetry of the barrier we plot with a red dotted line the barrier from the final to the initial points, i.e., the CW rotation

To conclude this section, we have shown that the NEB method within DFT is able to qualitative reproduce the asymmetric energy barrier of the rotation of C_2H_2 on PdGa:A$(\bar{1}\bar{1}\bar{1})$Pd$_3$. However, our most accurate values for the barrier height ($\Delta E_{B,\text{ZPE}} = 46$ meV), and barrier asymmetry ($R_{\text{asym}} = 1.06$) are far from the experimental ones ($\Delta E_B = 27.5 \pm 7.1$ meV, and $1.25 < R_{\text{asym}} < 1.5$).

5 DMBI Derivatives on Au(111)

In this chapter we explore a family of molecules which, after deposition on a Au(111) surface, become chiral. Depending on the adsorption on the surface they form a molecule-rotor when chemisorbed, or a drivable object ("nanocar") when physisorbed. This family of molecules is formed by 2-(2-Methoxyphenyl)-1, 3-dimethyl-1H-benzoimidazol-3-ium iodide (*o*-MeO-DMBI-I) and other DMBI derivatives, which were originally synthesized as air-stable *n*-type dopants in organic electronic devices [47, 48].

o-MeO-DMBI-I was sublimated at 490 K under UHV conditions and deposited on Au(111) [6]. After deposition the molecule was chemisorbed. Using DFT calculations, different options were considered and compared with the experimental data. In particular, previous studies [49] had shown that the molecule can undergo an intramolecular reaction where the O-Me bond can be cleaved, reacting with the iodine atom to form methyl iodide (MeI). After the elimination of MeI, zwitterionic 2-(1, 3-dimethyl-1H-benzo[*d*]imidazol-3-ium-2-yl)-phenolate (DMBI-P) is formed. DFT results showed that the zwitterionic DMBI-P was the compound observed on the surface [12]. Additional chemical studies in gas phase confirmed that the molecular dissociation takes places at the experimental conditions used in the sublimation of *o*-MeO-DMBI-I [12].

Following the results for DMBI-P, a second member of the family was synthesized. It was a zwitterionic phenolate terminated in a longer naphthyl, namely 2-(1, 3-dimethyl-1H-naphtho[2, 3-d]imidazol-3-ium-2-yl) phenolate (DMNI-P). This compound was thermally deposited ($T_{evap} = 498$ K) under UHV conditions on the Au(111) surface kept at room temperature [13]. Depending on the evaporation time different coverages could be achieved. At lower coverages only single molecules were present, but as the coverage was increased dimers started to appear, becoming dominant at 0.2 ML [13].

DMNI-P isolated molecules on the surface looked similar to DMBI-P, but slightly longer. Both molecules were chiral on the surface, showing left and right orientations which could be distinguished in the topographic STM images. They were chemisorbed on the surface, forming strong bonds. DMBI-P could not be moved laterally neither by voltage pulses nor by lateral manipulation [12]. In the case of DMNI-P strong voltage pulses (>1.5 V) or lateral manipulation were only able to flip the molecule to the opposite side of the bond without changing the anchoring point. This process changed the chirality of the molecule. DFT calculations showed that the flipping mechanism consisted in the rotation of the phenolate around the O atom, while the rest of the molecule shifted laterally [13].

Both chiral molecules behaved as molecule-rotors on Au(111). At $T = 5$ K, applying a pulse on a particular region of the molecules it was possible to induce a one-step rotation around the anchoring point. The STM tip height was relatively far ($4 - 5$ Å) to avoid direct mechanical interaction. A typical voltage pulse of 0.5 V during 10 s was able to rotate a molecule 60° CW or CCW, depending on the chirality. Changing the voltage sign or the chirality of the molecule inverted the sense of the rotation from CW to CCW or vice versa. The directionality of the rotation was very high: around 99% of the attempts with the same voltage sign on molecules with the same chirality showed rotations in the same direction. While for DMNI-P 99% of the times the rotation was of 60°, for DMBI-P in 95% of the cases the rotation was of 60°, while in 4% of the cases the molecule rotated in the same direction, but to an intermediate position. Consecutive pulses were able to rotate the molecule 360°, recovering the initial position. The pulse had to be applied on a region which did not coincide with the anchoring point, so a reposition of the tip was necessary after each individual rotation step. In a full rotation the molecule visited six stable positions corresponding to the sixfold symmetry of Au(111). By superimposing the STM images of the six stable positions in a full rotation the center of rotation could be experimentally determined. This confirmed the existence of a unique anchoring point for the molecule.

We performed DFT calculations in order to determine the ground state adsorption geometry of DMBI-P and DMNI-P on Au(111). In both cases it was shown that in the most stable position the molecule lies along the $(1\bar{1}0)$ direction of the surface. A 60° rotation of the molecule in the calculation leads to a configuration with the same energy within the error of the calculation. The anchoring point is shown to be the phenolate oxygen atom bound to a surface gold atom. The oxygen atom lies

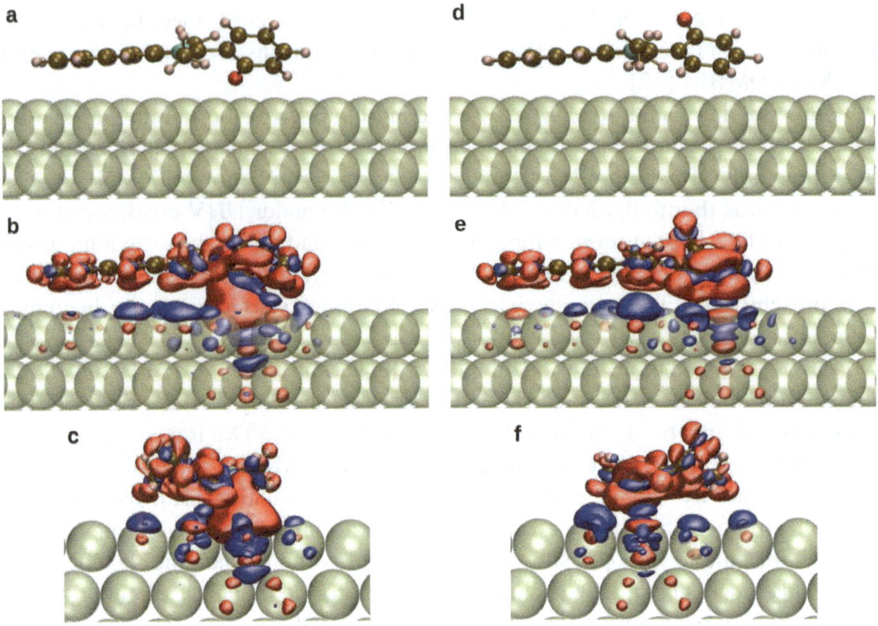

Fig. 7 Ball and stick models and induced charge density $\Delta\rho$ for DMNI-P on Au(111) with the oxygen atom in contact with the surface (**a–c**) and in the "up" position (**d–f**). (**a**, **d**) show the ball and stick models (brown, red, cyan, and pink balls represent C, O, N and H atoms, respectively. Big balls represent Au atoms). Middle and bottom panels represent the induced charge density in the X-Z plane (**b**, **e**) and Y-Z plane (**c**, **f**). Red (blue) colors indicate accumulation (depletion) of electrons

on top of the gold atom, but slightly off. The anchoring point and orientation is the same for both molecules, indicating that it is controlled by the phenolate end of the molecule. The O-Au distance is around 2.3 Å, compatible with a chemical O-Au bond. The phenolate forms a dihedral angle of 41° with the backbone of the molecule. We calculated the charge transfer by performing Bader analyses, finding that DMBI-P transfers 0.24 e- to the surface (0.26 e- for DMNI-P). In order to have more information about the O-Au bond we have computed the induced charge density $\Delta\rho$, which shows how the charge density is redistributed after the adsorption of the molecule on the surface. $\Delta\rho$ is defined as

$$\Delta\rho = \rho(\mathrm{mol@Au111}) - \rho(\mathrm{mol}) - \rho(\mathrm{Au111}),$$

where ρ(mol@Au111) is the charge density of the adsorbed molecule (DMBI-P or DMNI-P) at the surface, ρ(mol) is the charge density of the molecule, and ρ(Au111) is the charge density of Au(111), the last two quantities calculated separately, but keeping the same geometry of the adsorbed case. The results for DMNI-P are shown in the left side of Fig. 7. The position of the O atom can be seen in Fig. 7a. Around that position and connecting with the surface there is a strong charge accumulation (in red), which can be observed in both lateral views of Fig. 7b and c. An identical result is obtained for DMBI-P (not shown). This charge accumulation is due to the chemical bonding between O and the Au atom below.

The rotation of the molecules can be further analyzed by gradually increasing the temperature of the experiment. At around $T = 10$ K the rotation started to be thermally induced. When the temperature reached $T = 20$ K for DMBI-P [12] ($T =$ 29 K for DMNI-P [14]) the molecule rotates much faster than the scanning time for the STM image, resulting in a hexagonal image formed by the superposition of the six stable orientations of the molecule on the sixfold symmetric surface. The plot of the rate of the thermal induced rotations versus temperature showed an Arrhenius behavior, from which the barrier heights ΔE_B can be extracted. For DMBI-P the height was $\Delta E_B = 5 \pm 2$ meV [12], while for DMNI-P it was $\Delta E_B = 29 \pm 5$ meV [14]. The barrier is higher for the longer DMNI-P molecule due to the stronger interaction with the surface. In our calculations we computed a binding energy of 2.30 eV for DMBI-P, which is increased to 2.71 eV for DMNI-P. As we have seen, the behavior of the phenolate end of the molecule is identical in both cases, so the difference is due to the naphtyl end of DMNI-P, which shows a stronger vdW binding with the surface.

As indicated previously, when increasing the coverage for DMNI-P, dimers started to appear until they became dominant at a coverage of 0.2 ML. The dimers are formed by identical DMNI-P molecules. They can be separated by voltage pulses applied directly on the dimer, or applied as far as 8 nm away from it if the pulse is intense enough [13]. After the splitting, at least one of the single molecules didn't show an anchoring point. Instead it could be translated laterally in a controlled way, acting as a nanocar. Around 60% of the separated dimers resulted in a rotor and a nanocar, while the remaining 40% resulted in two nanocars. By applying positive voltage pulses the nanocar is attracted towards the tip with high precision. For high positive voltages the molecule can move up to 2 nm ($V > 1.5$ V). For negative pulses, the molecule moves away from the tip in random directions. The behavior of the nanocar is independent of its chirality, in contrast to chiral-dependent rotation of the molecule.

The properties of the nanocar could be due to different reasons. The dimers could involve the presence of additional Au adatoms, as it was shown for supramolecular assemblies on Au(111) [50]. After separation of the dimer, the resulting molecules could be attached to an adatom (nanocars), or could be anchored to surface (rotors). The oxygen in the phenolate could also be passivated by an extra hydrogen atom which would preclude the bonding to the surface of the nanocars. A third option, which was considered the most likely in [13], was traced back to different geometrical conformations of the molecule on the surface. DFT calculations find a second stable configuration of DMNI-P on the surface, where the phenolate is rotated with the oxygen in an "up" position, away from the surface, as seen Fig. 7d. This configuration is 0.22 eV less stable than the one with the oxygen bonded to the surface, but would be stabilized by the process of splitting the dimer, and by the relative high energy barrier to fall back to the ground state (estimated as 0.4 eV in [13]). This configuration is physisorbed to the surface, and its binding energy is essentially due

Fig. 8 Calculated vibrational modes of DMBI-P on Au(111) (black) and DMNI-P on Au(111) with the oxygen in contact with the surface (red) and in the "up" position (green)

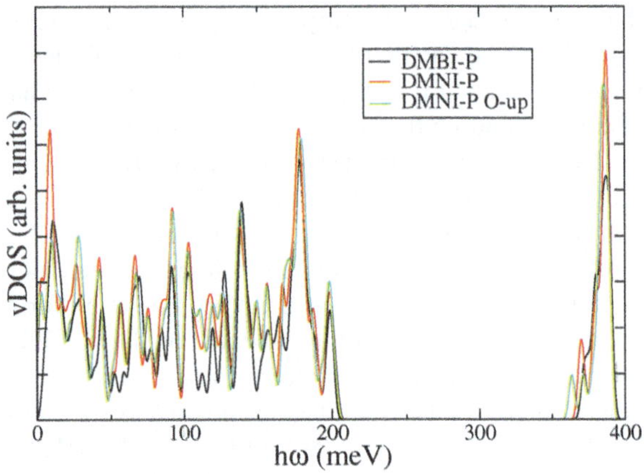

to vdW interactions. The difference of the bonding between both configurations can be observed by comparing the induced charge densities $\Delta\rho$ in Fig. 7. $\Delta\rho$ for DMNI-P with the O atom in the "up" position is shown in Fig. 7d–f. For this geometry there is no induced charge density around the O atom, in contrast with the strong charge accumulation for the chemisorbed geometry.

Extra information of the behavior of the molecules can be extracted by measuring the action spectra, plotting the yield (motion probability per electron) versus bias voltage. This was done for DMBI-P rotors [12] and for DMNI-P rotors and nanocars [13]. Action spectra for the rotating molecules were fitted by the theory in [51], finding the same threshold of 370 meV for both DMBI-P and DMNI-P. This value corresponds to the minimum excitation energy needed to rotate the molecule. The most straightforward excitations of the molecules are vibrational, so we have calculated the vibrational modes of the molecules on the surface by DFT. The corresponding vibrational density of states (vDOS) is shown in Fig. 8. For DMBI-P, the vDOS shows a gap between the end of the C-H rotational modes at 199 meV until the onset of the C-H stretch modes at 371 meV (between 199 meV and 369 eV for DMNI-P). The last value perfectly agrees with the experimental threshold of 370 meV. Therefore, we can conclude that the rotation of the molecule is activated by the efficient injection of energy by the C-H stretch modes. In particular, the first mode to appear corresponds to a C-H stretch mode of the methyl group at the opposite side of the O-Au bond. In the case of the DMNI-P nanocar, the action spectrum can be fitted with a threshold voltage of 165 meV [13]. This value indicates that the translation of the physisorbed molecule can be activated by lower energies for the translation than for the rotations. Comparing to the calculated vibrational modes of the nanocar (green curve in Fig. 8), 165 meV would correspond to a mixture of C-C and C-N stretch modes between 155 and 167 meV.

The next step was to calculate the energy barriers using the NEB method. We started by carefully determining the initial and final configurations rotated by 60° for DMBI-P and DMNI-P. Running preliminary NEB calculations showed the pres-

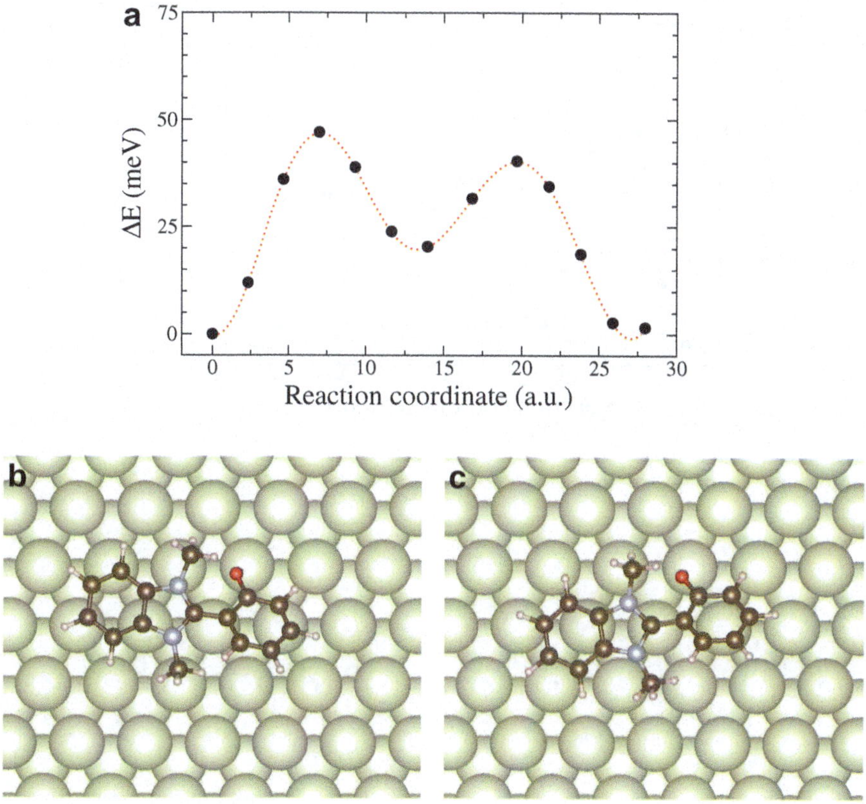

Fig. 9 **a** Energy barrier for a 60° rotation of DMBI-P on Au(111). The line is a cubic spline fitted to the data. Ball and stick models for the ground state (**b**) and the transition state (**c**). Brown, red, cyan, and pink balls represent C, O, N and H atoms, respectively. Big balls represent Au atoms

ence of intermediate local minima at around 30° for DMBI-P and around 20° and 40° for DMNI-P. Experimentally, 4% of rotations showed an intermediate position smaller than 60° for DMBI-P [12], which agrees with the presence of an intermediate local minimum. However, the same was not experimentally observed for DMNI-P. Next we optimized the local minima and divided the total NEB calculation into two NEB runs with 5 images each for DMBI-P, and in three NEB runs with 5 images each for DMNI-P. The results are shown for DMBI-P in Fig. 9 and for DMNI-P in Fig. 10. We immediately see that the barriers are asymmetric, although due to the presence of several TSs their analysis is more complicated than for C_2H_2 on PdGa (Sect. 4). For both DMBI-P and DMNI-P we see first the higher TS, followed by one or two shallower TSs. The higher TS shows heights of ΔE_B(DMBI-P) = 47 meV, and of ΔE_B(DMNI-P) = 64 meV. The inclusion of ZPE corrections reduces theses values to $\Delta E_{B,\mathrm{ZPE}}$(DMBI-P) = 45 meV and $\Delta E_{B,\mathrm{ZPE}}$(DMNI-P) = 60 meV, respectively. These values are bigger than the experi-

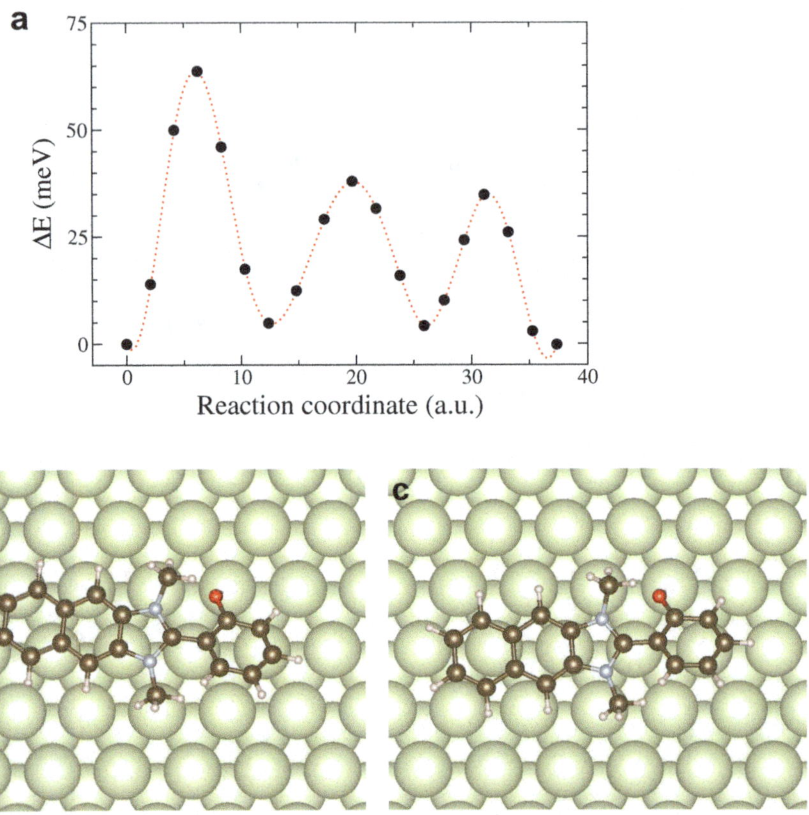

Fig. 10 **a** Energy barrier for a 60° rotation of DMNI-P on Au(111). The line is a cubic spline fitted to the data. Ball and stick models for the ground state (**b**) and the transition state (**c**). Brown, red, cyan, and pink balls represent C, O, N and H atoms, respectively. Big balls represent Au atoms

mental results of ΔE_B(DMBI-P) $= 5 \pm 2$ meV, and ΔE_B(DMNI-P) $= 29 \pm 5$ meV. However, the calculations correctly give a higher value of the barrier height for DMNI-P, due to the stronger vdW binding of the rotating part in the longer molecule.

Examination of the NEB results allow us to extract more information. In Figs. 9 and 10 we show the geometry of the ground state in (b) and of the transition state in (c). It can be seen that in the GS the phenyl or naphthyl groups lie on bridge or hollow positions of Au(111). In the TS these groups lie mainly on the top positions. Similar results hold for every local minima and TSs. Further analysis of the calculated NEB trajectory shows that the axle of rotation is best described as an axis perpendicular to the surface and going through the surface Au atom bonded to the O atom.

To conclude this section, we have seen that DFT calculations allow us to describe the unidirectional rotation of DMBI derivatives on Au(111). We have been able to describe the chemisorption of DMBI-P and DMNI-P due to the formation of a Au-O bond. We have shown that the rotation is due to the injection of energy into the molecule by inelastic excitations of the C-H stretch modes. Finally, we have showed that the NEB method correctly produces asymmetric barriers but, as for C_2H_2 on PdGa, the barrier heights are too high compared to the experiment.

6 Conclusions and Outlook

We have studied the asymmetric rotational energy barriers of molecule-rotors by using the NEB method within DFT. First we have assessed the ability of the method to produce symmetric barriers for the achiral DBBP in the gas phase. Then we have applied the method to C_2H_2 on the chiral surface PdGa:A$(\bar{1}\bar{1}\bar{1})$Pd$_3$, finding that NEB is able to find asymmetric barriers which qualitatively explain the directionality of the rotation, but which are too high compared with the experimental values. In the last part we have studied DMBI derivatives on Au(111). We have found that the barriers for these systems are asymmetric and depend on the details of the molecule. In particular, longer molecules have a stronger vdW interaction with the surface, which increases the height of the barrier. Compared with the experimental results the barriers are also too high. The tendency of the NEB method within DFT to provide too high barriers has been shown in previous studies [4].

Although we have proven that the NEB method within DFT is able to provide qualitative results in agreement with the experiment, quantitative results are still lacking. A way of improvement would be to have a better description of the molecule-surface interaction. For example, a better description of the vdW forces could be beneficial. It has also been shown that the interaction between the STM tip and the molecule can modify the PES [46]. Therefore the inclusion of the STM tip (or at least of its main effects) in the calculation could also be important for the correct description of unidirectional rotations of molecule-rotors. These possibilities will be explored in future works.

Acknowledgements We acknowledge financial support from the European Union projects MEMO 766864 and ESiM 101046364 and from the Spanish State Research Agency grant (Project No. PID2021-127917NB-I00) funded by MCIN/AEI/10.13039/501100011033 and by "ERDF a way of making Europe". We are grateful for the computer resources at Finisterrae III and the technical support provided by CESGA. Views and opinions expressed are however those of the authors only and do not necessarily reflect those of the European Union. Neither the European Union nor the granting authority can be held responsible for them.

References

1. Leigh, D.A., Wong, J.K.Y., Dehez, F., Zerbetto, F.: Nature **424**(6945), 174 (2003). https://doi.org/10.1038/nature01758
2. van Delden, R.A., ter Wiel, M.K.J., Pollard, M.M., Vicario, J., Koumura, N., Feringa, B.L.: Nature **437**(7063), 1337 (2005). https://doi.org/10.1038/nature04127

3. Tierney, H.L., Murphy, C.J., Jewell, A.D., Baber, A.E., Iski, E.V., Khodaverdian, H.Y., McGuire, A.F., Klebanov, N., Sykes, E.C.H.: Nat. Nanotechnol. **6**(10), 625 (2011). https://doi.org/10.1038/nnano.2011.142
4. Tierney, H.L., Han, J.W., Jewell, A.D., Iski, E.V., Baber, A.E., Sholl, D.S., Sykes, E.C.H.: J. Phys. Chem. C **115**(4), 897 (2011). https://doi.org/10.1021/jp1026702
5. Perera, U.G.E., Ample, F., Kersell, H., Zhang, Y., Vives, G., Echeverria, J., Grisolia, M., Rapenne, G., Joachim, C., Hla, S.W.: Nat. Nanotechnol. **8**(1), 46 (2013). https://doi.org/10.1038/nnano.2012.218
6. Eisenhut, F., Meyer, J., Krüger, J., Ohmann, R., Cuniberti, G., Moresco, F.: Surf. Sci. **678**, 177 (2018). https://doi.org/10.1016/j.susc.2018.05.003
7. Simpson, G.J., García-López, V., Daniel Boese, A., Tour, J.M., Grill, L.: Nat. Commun. **10**(1), 4631 (2019). https://doi.org/10.1038/s41467-019-12605-8
8. Zhang, Y., Calupitan, J.P., Rojas, T., Tumbleson, R., Erbland, G., Kammerer, C., Ajayi, T.M., Wang, S., Curtiss, L.A., Ngo, A.T., Ulloa, S.E., Rapenne, G., Hla, S.W.: Nat. Commun. **10**(1), 3742 (2019). https://doi.org/10.1038/s41467-019-11737-1
9. Jasper-Toennies, T., Gruber, M., Johannsen, S., Frederiksen, T., Garcia-Lekue, A., Jäkel, T., Roehricht, F., Herges, R., Berndt, R.: ACS Nano **14**(4), 3907 (2020). https://doi.org/10.1021/acsnano.0c00029
10. Ren, J., Freitag, M., Schwermann, C., Bakker, A., Amirjalayer, S., Rühling, A., Gao, H.Y., Doltsinis, N.L., Glorius, F., Fuchs, H.: Nano Lett. **20**(8), 5922 (2020). https://doi.org/10.1021/acs.nanolett.0c01884
11. Stolz, S., Gröning, O., Prinz, J., Brune, H., Widmer, R.: Proc. Natl. Acad. Sci. **117**(26), 14838 (2020). https://doi.org/10.1073/pnas.1918654117
12. Eisenhut, F., Kühne, T., Monsalve, J., Srivastava, S., Ryndyk, D.A., Cuniberti, G., Aiboudi, O., Lissel, F., Zobač, V., Robles, R., Lorente, N., Joachim, C., Moresco, F.: Nanoscale **13**(38), 16077 (2021). https://doi.org/10.1039/D1NR04583K
13. Au-Yeung, K.H., Sarkar, S., Kühne, T., Aiboudi, O., Ryndyk, D.A., Robles, R., Lorente, N., Lissel, F., Joachim, C., Moresco, F.: ACS Nano **17**(3), 3128 (2023). https://doi.org/10.1021/acsnano.2c12128
14. Au-Yeung, K.H., Sarkar, S., Kühne, T., Aiboudi, O., Ryndyk, D.A., Robles, R., Lissel, F., Lorente, N., Joachim, C., Moresco, F.: J. Phys. Chem. C **127**(34), 16989 (2023). https://doi.org/10.1021/acs.jpcc.3c04990
15. Tolman, R.C.: Proc. Natl. Acad. Sci. **11**(7), 436 (1925). https://doi.org/10.1073/pnas.11.7.436
16. Ernst, K.H.: Phys. Status Solidi (b) **249**(11), 2057 (2012). https://doi.org/10.1002/pssb.201248188
17. Kresse, G., Furthmüller, J.: Comput. Mater. Sci. **6**(1), 15 (1996). https://doi.org/10.1016/0927-0256(96)00008-0
18. Kresse, G., Joubert, D.: Phys. Rev. B **59**(3), 1758 (1999). https://doi.org/10.1103/PhysRevB.59.1758
19. Perdew, J.P., Burke, K., Ernzerhof, M.: Phys. Rev. Lett. **77**(18), 3865 (1996). https://doi.org/10.1103/PhysRevLett.77.3865
20. Stephens, P.J., Devlin, F.J., Chabalowski, C.F., Frisch, M.J.: J. Phys. Chem. **98**(45), 11623 (1994). https://doi.org/10.1021/j100096a001
21. Tkatchenko, A., Scheffler, M.: Phys. Rev. Lett. **102**(7), 073005 (2009). https://doi.org/10.1103/PhysRevLett.102.073005
22. Ruiz, V.G., Liu, W., Zojer, E., Scheffler, M., Tkatchenko, A.: Phys. Rev. Lett. **108**(14), 146103 (2012). https://doi.org/10.1103/PhysRevLett.108.146103
23. Grimme, S., Antony, J., Ehrlich, S., Krieg, H.: J. Chem. Phys. **132**(15), 154104 (2010). https://doi.org/10.1063/1.3382344
24. Grimme, S., Ehrlich, S., Goerigk, L.: J. Comput. Chem. **32**(7), 1456 (2011). https://doi.org/10.1002/jcc.21759
25. Jónsson, H., Mills, G., Jacobsen, K.: In: Berne, B., Ciccotti, G., Jacobsen, K. (eds.) Classical and Quantum Dynamics in Condensed Phase Simulations, p. 385. World Scientific, Singapore (1998)

26. Henkelman, G., Jónsson, H.: J. Chem. Phys. **113**(22), 9978 (2000). https://doi.org/10.1063/1.1323224
27. Henkelman, G., Uberuaga, B.P., Jónsson, H.: J. Chem. Phys. **113**(22), 9901 (2000). https://doi.org/10.1063/1.1329672
28. Tang, W., Sanville, E., Henkelman, G.: J. Phys. Condens. Matter **21**(8), 084204 (2009). https://doi.org/10.1088/0953-8984/21/8/084204
29. Tersoff, J., Hamann, D.R.: Phys. Rev. B **31**(2), 805 (1985). https://doi.org/10.1103/PhysRevB.31.805
30. Bocquet, M.L., Lesnard, H., Monturet, S., Lorente, N.: In: Santen, R.A.V., Sautet, P. (eds.) Computational Methods in Catalysis and Materials Science, pp. 199–219. Wiley-VCH Verlag GmbH & Co. KGaA (2009)
31. Lorente, N., Robles, R.: STMpw v1.0b2 (Zenodo) (2019). https://doi.org/10.5281/zenodo.3581159
32. Momma, K., Izumi, F.: J. Appl. Crystallogr. **44**(6), 1272 (2011). https://doi.org/10.1107/S0021889811038970
33. Humphrey, W., Dalke, A., Schulten, K.: J. Mol. Graph. **14**(1), 33 (1996). https://doi.org/10.1016/0263-7855(96)00018-5
34. Almenningen, A., Bastiansen, O., Fernholt, L., Cyvin, B.N., Cyvin, S.J., Samdal, S.: J. Mol. Struct. **128**(1), 59 (1985). https://doi.org/10.1016/0022-2860(85)85041-9
35. MacNeil, D.D., Decken, A.: Acta Crystallogr. A **C55**(4), 628 (1999). https://doi.org/10.1107/S0108270198015339
36. Grein, F.: J. Phys. Chem. A **106**(15), 3823 (2002). https://doi.org/10.1021/jp0122124
37. Prinz, J., Gaspari, R., Pignedoli, C.A., Vogt, J., Gille, P., Armbrüster, M., Brune, H., Gröning, O., Passerone, D., Widmer, R.: Angew. Chem. Int. Ed. **51**(37), 9339 (2012). https://doi.org/10.1002/anie.201203787
38. Rosenthal, D., Widmer, R., Wagner, R., Gille, P., Armbrüster, M., Grin, Y., Schlögl, R., Gröning, O.: Langmuir **28**(17), 6848 (2012). https://doi.org/10.1021/la2050509
39. Krajčí, M., Hafner, J.: J. Chem. Phys. **138**(12), 124703 (2013). https://doi.org/10.1063/1.4795435
40. Bechthold, P., Ardhengi, J.S., Juan, A., González, E.A., Jasen, P.V.: Appl. Surf. Sci. **315**, 467 (2014). https://doi.org/10.1016/j.apsusc.2014.01.074
41. Bechthold, P., Ardenghi, J.S., Nagel, O., Juan, A., González, E.A., Jasen, P.V.: Int. J. Hydrogen Energy **39**(5), 2093 (2014). https://doi.org/10.1016/j.ijhydene.2013.11.077
42. Prinz, J., Pignedoli, C.A., Stöckl, Q.S., Armbrüster, M., Brune, H., Gröning, O., Widmer, R., Passerone, D.: J. Am. Chem. Soc. **136**(33), 11792 (2014). https://doi.org/10.1021/ja505936b
43. Prinz, J., Gaspari, R., Stöckl, Q.S., Gille, P., Armbrüster, M., Brune, H., Gröning, O., Pignedoli, C.A., Passerone, D., Widmer, R.: J. Phys. Chem. C **118**(23), 12260 (2014). https://doi.org/10.1021/jp501584f
44. Prinz, J., Gröning, O., Brune, H., Widmer, R.: Angew. Chem. Int. Ed. **54**(13), 3902 (2015). https://doi.org/10.1002/anie.201410107
45. Yakutovich, A.V., Hoja, J., Passerone, D., Tkatchenko, A., Pignedoli, C.A.: J. Am. Chem. Soc. **140**(4), 1401 (2018). https://doi.org/10.1021/jacs.7b10980
46. Lu, H.L., Cao, Y., Qi, J., Bakker, A., Strassert, C.A., Lin, X., Ernst, K.H., Du, S., Fuchs, H., Gao, H.J.: Nano Lett. **18**(8), 4704 (2018). https://doi.org/10.1021/acs.nanolett.8b01019
47. Wei, P., Menke, T., Naab, B.D., Leo, K., Riede, M., Bao, Z.: J. Am. Chem. Soc. **134**(9), 3999 (2012). https://doi.org/10.1021/ja211382x
48. Lüssem, B., Keum, C.M., Kasemann, D., Naab, B., Bao, Z., Leo, K.: Chem. Rev. **116**(22), 13714 (2016). https://doi.org/10.1021/acs.chemrev.6b00329
49. Schwarze, M., Naab, B.D., Tietze, M.L., Scholz, R., Pahner, P., Bussolotti, F., Kera, S., Kasemann, D., Bao, Z., Leo, K.: ACS Appl. Mater. Interfaces **10**(1), 1340 (2018). https://doi.org/10.1021/acsami.7b14034
50. Robles, R., Zobač, V., Yeung, K.H.A., Moresco, F., Joachim, C., Lorente, N.: Phys. Chem. Chem. Phys. **22**, 15208 (2020). https://doi.org/10.1039/D0CP01657H
51. Kim, Y., Motobayashi, K., Frederiksen, T., Ueba, H., Kawai, M.: Prog. Surf. Sci. **90**(2), 85 (2015). https://doi.org/10.1016/j.progsurf.2014.12.001

Driving a Single Chemisorbed Molecule-Rotor by Thermal Energy and Tunneling Electrons

Kwan Ho Au-Yeung, Suchetana Sarkar, Christian Joachim, and Francesca Moresco

Abstract This study investigates the unidirectional rotation of single molecule-rotors, crucial for nanoscale work. We compared two DMBI-based molecule-rotors on an Au(111) surface, exploring their one-way rotation induced by tunneling electrons and thermally-induced random rotation rates. By optimizing the parameters of temperature and tunneling current, we identified a narrow voltage range that enhances unidirectional rotation without inelastic tunneling effects. By the quantum mixing of the ground and excited states occurring at each tunneling electron transfer event through the molecule-rotor, the thermal energy can be projected in part on the excited states of the molecule-rotor to enhance the one-way rotation effect in a bias voltage range where inelastic tunneling effects are nearly absent.

1 Introduction

Since more than two decades, the motion of a single molecule adsorbed on the surface of a solid can be induced and probed using a scanning tunnelling microscope (STM) [1–18]. The semi-classical mechanical motion of an individual molecule spatially confined on an atomically clean metallic surface can be triggered by STM tip-induced manipulation, while its translation and/or rotation can be observed by STM imaging. Inspired by macroscopic machines, molecule-gears, rotors/motors, and nano-vehicles (Nanocar) have been recently investigated, which are often driven by completely different and sometimes not completely understood mechanisms [19–23]. One main goal of this new field of research is to produce useful work at the nanoscale by a single molecule adsorbed on a surface, developing design strategies for mechanics and energetics.

K. H. Au-Yeung · S. Sarkar · F. Moresco (✉)
Center for Advancing Electronics Dresden, TU Dresden, 01069 Dresden, Germany
e-mail: francesca.moresco@tu-dresden.de

C. Joachim
GNS & MANA Satellite, CEMES, CNRS, 29 rue J. Marvig, 31055 Toulouse, France

X. Bouju and C. Joachim (eds.), *Crossroad of Maxwell Demon*, Advances in Atom and Single Molecule Machines, https://doi.org/10.1007/978-3-031-57904-2_9

According to the second law of thermodynamics, one has to always "pay" for the energy needed to reduce the entropy of a given macroscopic machine. To achieve any distribution of energy other than the thermodynamic equilibrium inside the machine, external work must be performed. Only the intervention of a Maxwell's demon [24] could avoid this work. More interestingly, only the intervention of a demon could help to locally direct this energy inside the macroscopic machine without respecting any equipartition. Involving quantum instead of classical mechanics, single molecule machines provide new perspectives in this context. While the mechanics of a classical machine is determined by one single potential energy surface (PES), a quantum machine can take advantage from multiple potential energy surfaces, from ground to excited electronic states.

2 The Micro-Reversibility Principle

For a single molecule machine, unidirectional rotation represents a major challenge. When the machine delivers sufficient motive power, it is able to rotate itself and an external load [25]. However, thermal fluctuations coming from the background (e.g., solid surface, membrane, solution, or vacuum) only involve the electronic ground state PES, and thus cannot induce directed motion without violating "the principle of microscopic reversibility" [26]. This principle asserts that when a system is in thermal equilibrium, the rate of all microscopic processes, whether in forward or reverse directions, must be equal, thereby precluding the occurrence of a directed motion.

According to the energy equipartition theorem [27], thermal energy is evenly distributed among all mechanical degrees of freedom of the molecule machine supported on the surface of a solid maintained at a temperature T. If we represent the process by a set of given reaction coordinates, we obtain a potential energy profile on the ground state PES which is natively harmonic in all directions, thus leading to the principle of microscopic reversibility. Figure 1 illustrates the harmonicity of the potential energy curve with two local minima within the ground state, and the transition between two stable and one metastable mechanical configurations of the molecule. Thermal excitations (k_BT) captured by the machine from the background lead to random rotational movements if the amount of energy is enough to pass over the central saddle point of Fig. 1. In essence, rapid interconversions between the two local minima occur, driven by thermal contributions, making unidirectional rotation not possible in this scenario.

Studied in solution, a now well-known molecule from B. L. Feringa shed light on the dynamics behind the unidirectional rotation of molecular motors ("Feringa's motor") [28]. Powered by light, the molecules are derived from chiral overcrowded alkenes. They can undergo repetitive cycles of photochemical and thermal rearrangements leading in average to a directed motion. Density Functional Theory (DFT) calculations [29] delves into the Born–Oppenheimer like dynamics on the PESs underlying the photoisomerization and thermal helix inversion of the Feringa's

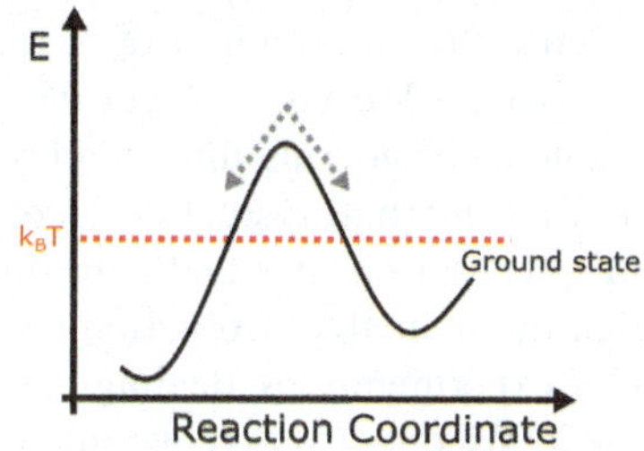

Fig. 1 Schematic potential energy profile in the simplest picture. Random rotation due to thermal excitation (k_BT)

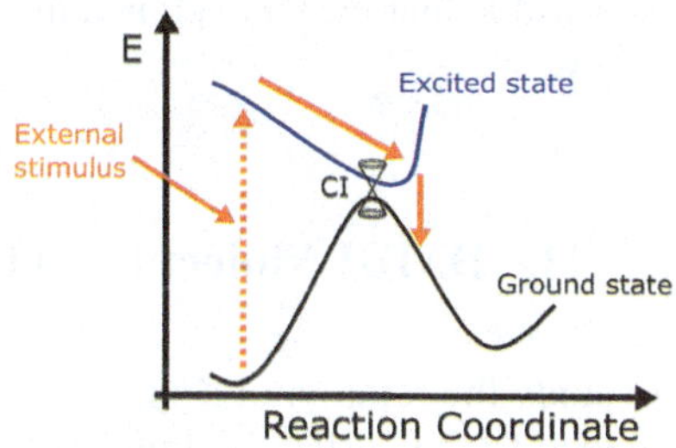

Fig. 2 Motor rotation dynamics on the ground state (lower) and excited state (upper) potential energy profiles in the simplest view. The profile of thermal helix inversion in the case of Feringa's motor is not shown here

motor. Figure 2 proposes a simplified explanation of this one-way step by step triggered rotation, highlighting a photoisomerization process, without displaying the more complex energy reaction path for the thermal helix inversion process.

The experimental results of the Feringa's motors clearly demonstrate the importance of the electronic excited states in the rotation process and the subsequent relaxation through a hyperspace funnel, known as a conical intersection (CI). It leads to the next local energy minimum along the reaction coordinate. CI occurs when two Born–Oppenheimer potential energy surfaces (for example, the ground and first excited state) intersect, forming in some cases a double cone crossing at the cone summit contact point. In this unique position, the molecule can easily pass between different PESs. Currently, this phenomenon is widely acknowledged for its pivotal role in nonadiabatic processes, such as physiological vision and light harvesting.

In a simplified version of the rotation process (Fig. 2), the rotating part of the molecular motor starts from a first stable potential energy minimum. It is then excited by light to undergo a vertical transition (orange arrows) in a Frank–Condon process reaching the excited PES. When the resulting intramolecular dynamics on this PES is reaching the CI point between the excited and ground state potential energy surfaces, the corresponding trajectory reaches the ground state PES at the CI point and continues down the PES to the next (meta-)stable local minimum. Such process allows to develop a unidirectional reaction path along the ground state PES. Here an asymmetric local topography of the excited PES is required. This asymmetry is indirectly related to the chemical structure of the molecular motor.

In this chapter, we discuss and compare the mechanical behavior of two single molecule rotors chemisorbed on the Au(111) surface, which can undergo unidirectional rotation induced by bias voltage pulses applied in the STM junction. We then discuss the contribution of the inelastic electron tunnelling (IET) process during the pulses. We also present the random rotation of the molecule while thermally excited, considering the interplay between electron tunnelling through the molecule and the thermal excitation from the Au(111) supporting surface. From the view of quantum physics and considering the experimental condition, one should note that the tunnelling excitation process is different from photochemistry in solution. Our experiments are dealing with one single adsorbed molecule during one experiment. We only average the multiple electron transfer events per second through the molecule-rotor, which defines the tunneling current intensity. It may also therefore be possible that the CI exploited in solution is not required in this case, as discussed below.

3 The DMBI Molecule and Its Derivative as Rotor Unit

To study the rotation of a single adsorbed molecule, the molecule must be strongly bonded to the surface to limit the possibility of lateral diffusion. Several possible anchoring strategies have been reported [23, 30–32], the anchoring interaction being also used as a robust rotation axle. In this chapter, we compare two single molecule-rotors having a similar chemical structure which can anchor on the Au(111) surface via a single chemical bond.

The 2-(2-Methoxyphenyl)-1,3-dimethyl-1H-benzoimidazol-3-ium iodide (o-OMe-DMBI-I) molecular precursor [33] and its derivatives [34] have been first synthesized for developing n-type dopants with high air stability for organic electronic applications. It was discovered later [35] that a decomposition reaction can occur during the doping process at high temperature, with the halide separation and the separation of a methyl group. In ultra-high vacuum (UHV), the o-OMe-DMBI-I precursor also undergoes a similar decomposition reaction during its sublimation at around 220 °C. Here, the methoxy group is cleaved and a methyl iodide is released leading to the zwitterionic phenolate DMBI-P compound [36] (Fig. 3a). On the Au(111) surface, experiments on DMBI derivatives were motivated by this zwitterionic nature and by its excellent unidirectional rotation induced by the application of STM voltage pulses.

A new DMBI-like molecule but with one extended benzene ring more (going from a benzimidazole moiety to a naphtho[2, 3-d]imidazole compound) was later designed and synthesized. In this case, the methoxy cleavage is superfluous and the rotating part is elongated from the phenyl (Fig. 3b) to the 2-(1, 3-dimethyl-1H-naphtho[2, 3-d]imidazol-3-ium-2-yl) phenolate (DMNI-P) compound. This extension was performed to increase the van der Waals (vdW) interaction between the naphtyl and the Au(111) as compared to the phenyl, with the aim of influencing the potential energy rotation barrier height of the molecule rotor.

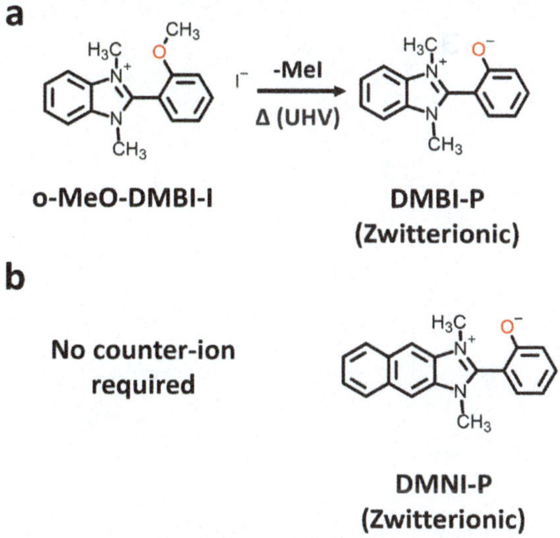

Fig. 3 DMBI and derivative as single molecule-rotors

4 Adsorption and Unidirectional Rotation of a Single DMBI-P Molecule on Au(111)

STM experiments on the DMBI molecules were performed after cooling the gold sample to T = 5 K. After the sublimation (at 217 °C) of the o-OMe-DMBI-I precursor on a clean Au(111) surface kept at room temperature, the representative large scale STM image of Fig. 4a shows a very low molecular coverage where single isolated molecules are visible at kink sites of the Au(111) herringbone surface reconstruction and on the fcc terraces of this surface. A close-up STM image of a single molecule (Fig. 4b) shows a pear-shaped appearance with a distinguished lobe, in good agreement with the chemical structure of a DMBI-P molecule.

To confirm the assignment, Density Functional Theory (DFT) calculations of the optimized adsorption geometry of the molecule on the surface have been performed (Fig. 4c) showing a tilted phenolate due to the flexible C–C bond connecting to the benzimidazole body of the molecule, which can be assigned to the higher lobe of the STM image. The widening of the structure roughly in the middle of the STM image is coming from the two methyl groups of the imidazole unit. According to the DFT geometry optimization, a molecular ratchet effect can be anticipated because a planarization of the molecular structure would lead to a lower energy barrier for rotation as compared to a large one after an increase of the dihedral angle between the phenolate and the benzimidazole chemical groups.

At T = 5 K, when the STM tip apex is positioned opposite to the oxygen atom of the phenolate and a bias voltage pulse V(t) of typically 400 mV is applied, a one-step clockwise (CW) rotation of 60° around the Au–O bond is obtained (Fig. 5a). This rotation event is observed by recording the sudden change in the relative tip height z(t) after the raising up of the bias voltage (Fig. 5b). A subsequent STM image is

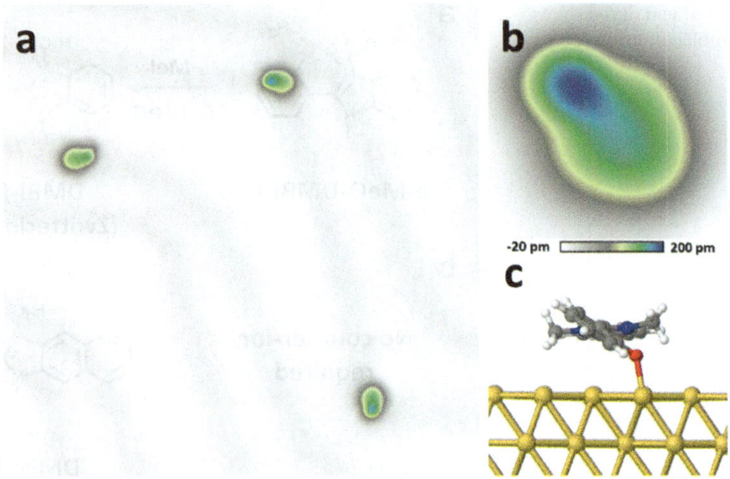

Fig. 4 Chemisorption of single DMBI-P molecules on Au(111). **a** Overview STM topography (20 nm × 20 nm, 0.2 V, 20 pA) after deposition of DMBI-P molecules. **b** Close-up STM image (1.8 nm × 1.8 nm, 0.2 V, 100 pA) of a single molecule. **c** Side view of the DFT calculated adsorption geometry of the molecule, showing the formation of the stable Au–O bond

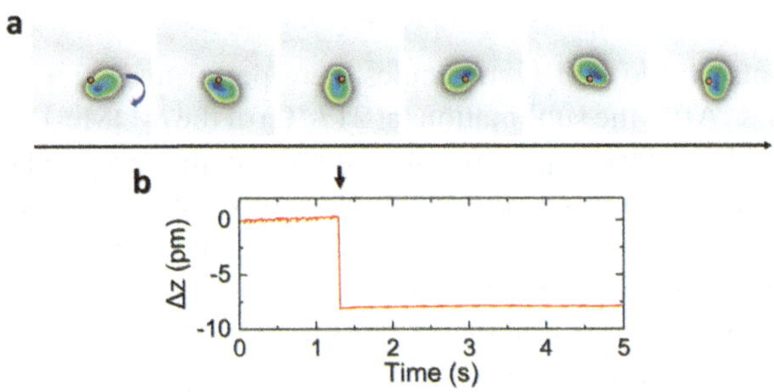

Fig. 5 Unidirectional rotation of DMBI-P induced by STM voltage pulses. **a** Sequence of STM images (5 nm × 5 nm, 200mV, 50 pA) of a DMBI-P molecule in the six adsorption orientations originating from the Au(111) symmetry. Topographies were taken consecutively after voltage pulses (V = 0.4 mV, I = 300 pA), and each pulse led to a clockwise 60° rotation. The tip is typically positioned above the phenolate moiety (the "head") of the molecule during the voltage pulses. The orange dots indicate that the molecule rotates at a fixed rotational center (Au–O bond). **b** An exemplary relative tip height over time (z-t) curve during a voltage pulse at constant current mode. A sudden change of the tip height can be observed, indicating a rotational event (black arrow)

used to certify the one-step rotation event. After five stepwise rotations, the molecule returns to its starting orientation after a final bias voltage pulse. It is important to note that the rotation can be CW or counter-clockwise (CCW) depending on the polarity of the applied bias voltage pulse and the molecular chirality on the surface. Such series of triggered rotation event maintaining the same anchoring point further confirms the strength of the Au–O bond.

5 The DMNI-P Single Molecule-Rotor

After the DMBI-P molecules, the new zwitterionic DMNI-P molecules were sublimated at 225 °C on a clean Au(111) surface kept at room temperature. In this case, no dissociation reaction is required to stabilize the anchoring point. The large scale STM image in Fig. 6a shows a very low coverage where the adsorption sites at kinks and terraces on the Au(111) surface are very similar to the case of DMBI-P.

The pear-shaped appearance from the STM image (Fig. 6b) features the same bent phenolate structure as confirmed by the DFT calculations (Fig. 6c). A larger lateral size of the molecule STM image is also observed due to the extension of the molecular structure from a phenyl to a naphtyl with widened sides coming from the electronic transparency of two methyl groups of the imidazole unit.

Identical to the experimental protocol described in Fig. 5 (T = 5 K), the STM tip apex is first positioned above the location of the carbon atom opposite to the oxygen atom of the molecule to induce rotation (Fig. 7a). Subsequently, a STM voltage pulse is applied in a constant current mode. The STM image recorded after the pulse (Fig. 7b) shows that in this case the molecule has rotated 60° CCW to reach the next stable station on the Au(111) surface. By applying a succession of pulses while following the orientation of the molecule, a complete CCW rotation of the molecule over its six stations of 60° can be observed, corresponding to all the six stable adsorption positions on a flat fcc terrace of Au(111).

The center of rotation (axle) can be experimentally determined by superimposing the STM images of six stations after a complete rotation cycle (Fig. 7c). Notice that

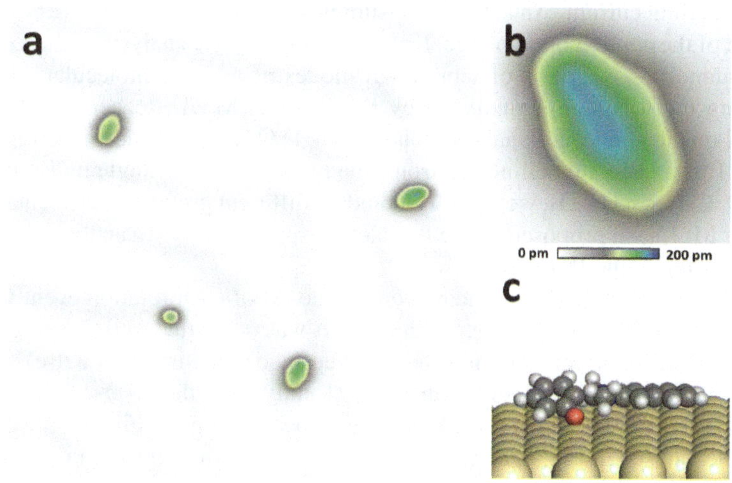

Fig. 6 Chemisorption of single DMNI-P molecules on Au(111). **a** Overview STM topography (40 nm × 40 nm, 0.2 V, 5 pA) after deposition of DMNI-P molecules. **b** Close-up STM image (1.8 nm × 1.8 nm, 0.2 V, 5 pA) of a single molecule. **c** Side view of the DFT calculated adsorption geometry of the molecule, showing the formation of the stable Au–O bond

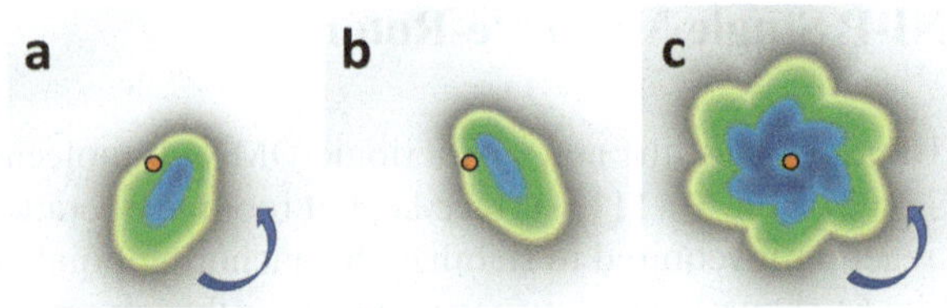

Fig. 7 Anchoring and rotation of a single DMNI-P. **a** STM image of the molecule anchored at the position of the orange dot. **b** STM image of the same DMNI-P rotor after a voltage pulse, showing a one-step rotation ($\Theta = 60°$; CCW). **c** By combining the images of the six stations rotating in one direction (CCW), the common anchoring point is identified (indicated by the orange dot)

by controlling the molecular coverage on Au(111), it was observed that a single DMNI-P molecule can be either chemisorbed with is oxygen atom looking towards the surface or physisorbed with is oxygen looking upwards. Only the chemisorbed conformation is demonstrating CW and CCW step by step rotation, the physisorbed molecule demonstrating a nanocar (lateral translation) behavior on Au(111) [37].

6 The Action Spectra for Rotation

We have further investigated the unidirectional rotation of both DMBI-P and DMNIP. Figure 8 shows the electronic yield, or "action spectra" [38] for DMBI-P and DMNI-P rotation and translation as a function of the bias voltage of the pulse at different fixed tunnelling current. An action spectrum is defined as a reaction rate (yield) as a function of the applied bias voltage. It allows a quantitative analysis of the excitation by the tunnelling electrons of vibrational modes of a single molecular adsorbate. Following the first study conducted by B. C. Stipe, M. A. Rezaei, and W. Ho in 1998 [1] on the non-directional rotation of a single O_2 molecule adsorbed on Pt(111) induced by inelastic tunnelling electrons, the motion of more single molecules [38], e.g., CO, C_2H_2 and C_2D_2, were soon studied by different groups experimentally. The group of M. Kawai proposed to use the "action spectroscopy" to analyze the motion corresponding to the IETS processes [39].

To plot the yield of Fig. 8, the response time Δt of each rotation event was first extracted from the z(t) signal (e.g., Fig. 5b). It was converted to the average rate $R = 1/\Delta t$ and the experimental yield was plotted by calculating $Y = R/I(e)$ with I(e) representing the tunnelling current intensity. Totally more than 600 rotational events for DMBI-P, and 250 events for DMNI-P have been plotted in Fig. 8.

As one can see in Fig. 8, at $V > 400$ mV the yield values for both molecules are very similar, resulting in a yield between 5×10^{-10} and 1×10^{-9} events/electron depending on the tunneling current intensity. A threshold voltage at around 400 mV is clearly observed for all cases, which can be explained by the onset of C–H stretch mode of the methyl group at the opposite side of the Au–O bond [36, 37]. These results demonstrate that DMBI-P and DMNI-P share the same spectral excitation of

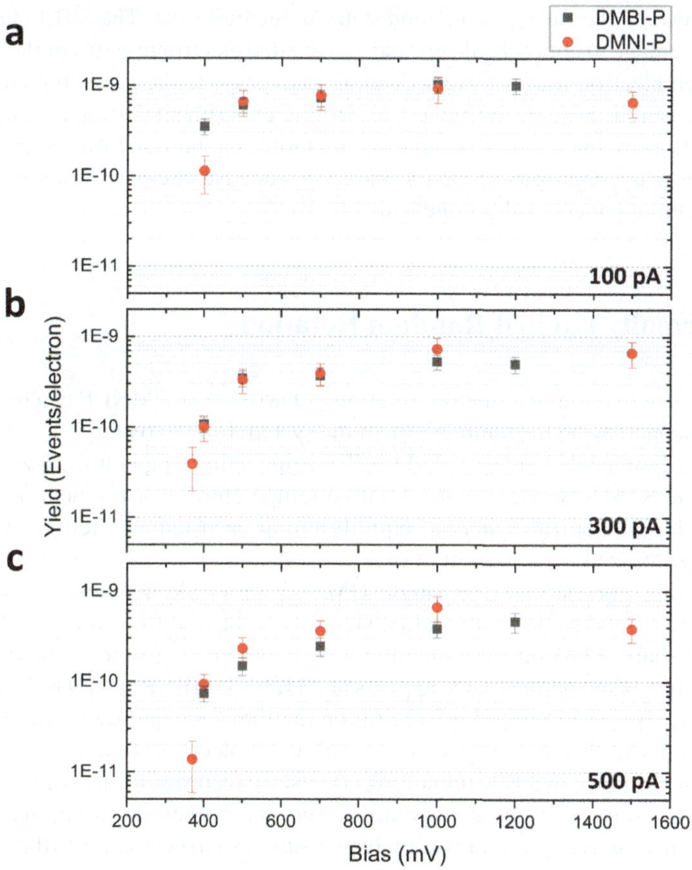

Fig. 8 Comparison of action spectra between DMBI-P and DMNI-P at different fixed tunnelling current setpoints: **a** 100 pA, **b** 300 pA, and **c** 500 pA. The black and red dots represent the datasets from DMBI-P and DMNI-P, respectively

the vibrational mode at the onset of the C–H stretch mode. However, one should note that the many random excitations of vibrational modes of a molecule acts similarly to a thermal excitation, where they are not sufficient to drive a one-step rotation of the DMBI-P or DMNI-P molecule around its axle. To do so, this energy must here be re-distributed intramolecularly in a collective mode direction corresponding to this rotation on the potential energy surfaces, ground and excited states.

According to the micro-reversibility principle, this energy redistribution is in competition with the intramolecular vibrational relaxation phenomenon which has the tendency to redistribute the inelastic energy initially provided to many accessible vibrational modes before releasing it towards the surface. Therefore, the observed step by step rotation is a consequence of an intramolecular competition between the rotation direction and the many directions of the vibrational modes on the ground state, but also on the electronic excited states of the molecule. The STM bias voltage has the advantage to entangle the ground and excited electronic states of the molecule during each electron transfer event, leading in average to the tunneling current that can be measured through the molecule. In this experimental situation, the micro-reversibility principle cannot be applied anymore, opening the question discussed below about the conditions needed to induce a stepwise one-way rotation using this quantum entanglement without inelastic effects.

7 Thermally Excited Random Rotation

The thermal activation of a random rotation of DMBI-P or DMNI-P can be achieved by stepwise increasing the temperature of the system formed by the Au(111) surface, the adsorbed molecule, and the STM tip. To regulate this temperature, a Zener diode fixed on the STM scanner was used with a temperature measurement accuracy of about 0.1 K. To attain the thermal equilibrium at each targeted temperature, each stage was stabilized for a minimum of 1 h.

Comparing Fig. 9a and c, a single DMNI-P molecule demands more thermal energy to reach the same rotational behavior as compared to a single DMBIP. The hexagonal shape STM image indicating a very fast visiting time of all six stations during the recording of one image appears at 20 K for DMBI-P and 29 K for DMNI-P, respectively. By positioning the STM tip on the molecule and by fixing its vertical distance to the surface in a way to not perturb the molecule rotation at low temperatures nor to introduce inelastic tunnelling effects, we recorded the I(t) signal (constant height mode) during the random rotation, and determined the rotational rate at a specific temperature. Quantitative analysis using an Arrhenius plot (the logarithm of a rotational rate against reciprocal of the temperature) reveals that ground state potential energy barrier of $E_{ex} = 5.3 \pm 2.5$ meV for DMBI-P and is $E_{ex} = 29.2 \pm 4.7$ meV for DMNI-P for one-way rotation (Fig. 9b and d). It is important to note that the exponential fits in Fig. 9b and d are based on a simplified model of a rigid rotor supposed to follow only the rotation direction on the potential energy surface without considering other possible trajectories on the ground state PES avoiding laterally the rotation saddle point. Furthermore, the experimental data were still recorded at very low temperatures, making it impossible to extrapolate the behavior of the rotor at much higher temperatures.

Such higher thermal energy required for a random rotation is the expected consequence of the extension of the chemical structure of DMNI-P as compared to DMBI-P, leading to a stronger vdW interactions with the Au(111) surface. As an indication, DFT calculations of the binding energy of those two molecules using vdW corrected DFT results in $E_{B,vdW} = 1.93$ eV for DMBI-P, and $E_{B,vdW} = 2.46$ eV for DMNI-P, respectively. Consistent with the principle of microscopic reversibility, those thermal excitation experiments confirm that a sole thermal heating of the DMNI-P or DMBI-P molecule-rotors does not cause any one-way rotation but instead leads to a random rotation process, taking place in their respective ground state electronic potential

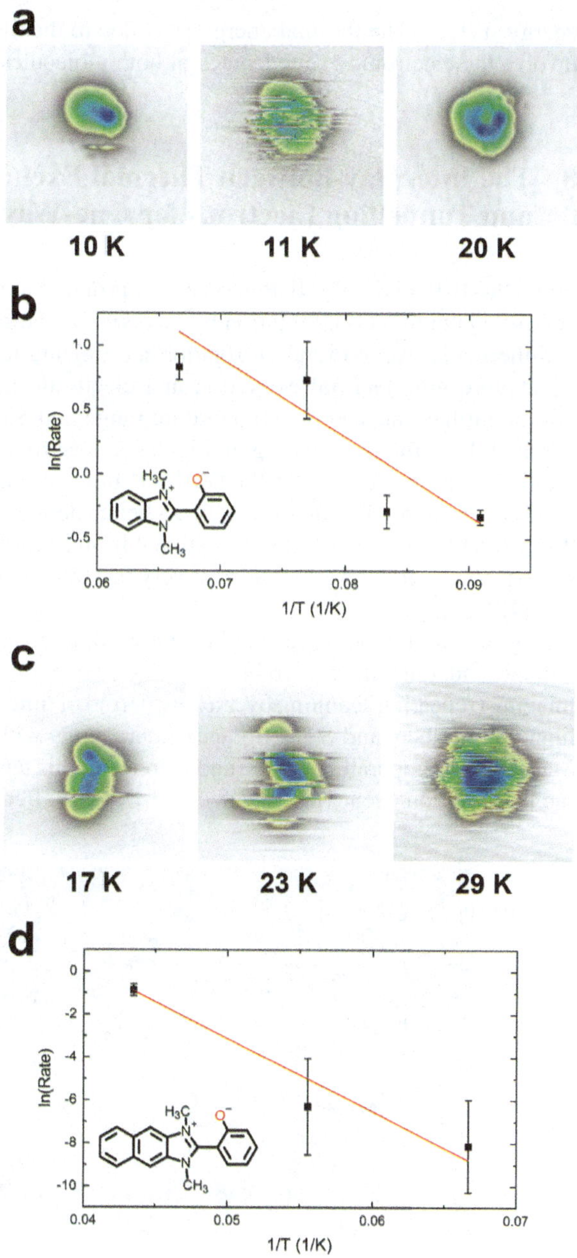

Fig. 9 Comparison of thermal excitation on DMBI-P and DMNI-P. **a**, **c** STM images (4 nm × 4 nm) recorded at different temperatures as indicated. Increasing the temperature leads to very fast random rotation visiting all six stations during the STM scan, and hence the molecular appearances show hexagonal shape when reaching **a** 20 K and **c** 29 K, respectively. **b**–**d** Arrhenius plots. The rotation rates are determined by recording the time interval between stations from the tunnelling current I(t) channel. The red lines are plotted by fitting $R = R_0 \exp(E_{ex}/k_B T)$, obtaining **c** $E_{ex} = (5.3 \pm 2.5)$ meV and **d** $E_{ex} = (29.2 \pm 4.7)$ meV, respectively

energy surface. The thermal energy provided to the molecule is not large enough to involve the electronic excited states of both molecules.

8 The Interplay Between Thermal Excitation and Tunnelling Electrons for One-Way Rotation

The DMBI-P and DMNI-P molecules provide a unique opportunity to study controlled unidirectional rotation by involving at the same time STM voltage pulses and thermally excited random rotation minimizing the inelastic tunnelling effects. To explore how thermal excitation and electronic tunnelling effect interplay, we have identified the specific temperature range and STM bias voltage pulses values in which both thermal heating and inelastic electronic tunnelling play a role in the rotation process. We focus on the DMNI-P molecule because of its wider temperature window for thermal excitation, hence a better-defined temperature setpoint to reach the different stages of a random rotational from a random hopping between only two stations to a visit of all six stations at very fast rate as compared to the STM scanning rate [40] (Fig. 9).

To find the temperature for the first random rotation event, we used very low voltage and tunnelling current (typically I = 5 pA and V = 10 mV) for STM imaging (repeating scanning over a long period of time). At these conditions, inelastic tunnelling effects and STM tip apex interactions with the molecule are negligible. Starting at a temperature of 5 K and increasing this temperature by steps of 0.5–1 K, the first rotation event was detected at T = 12 K after waiting 17 h (i.e., 0.06 event/

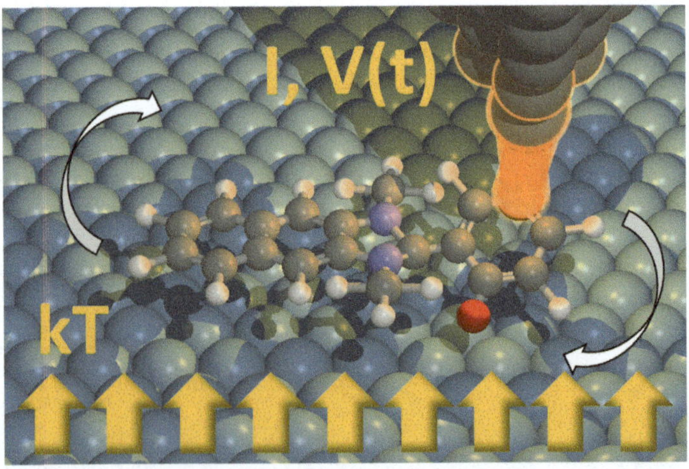

Fig. 10 Schematic drawing of DMNI-P rotor chemisorbed on Au(111). The work focuses on the interplay between inelastic tunneling (I, V(t)) and thermal excitations (kT)

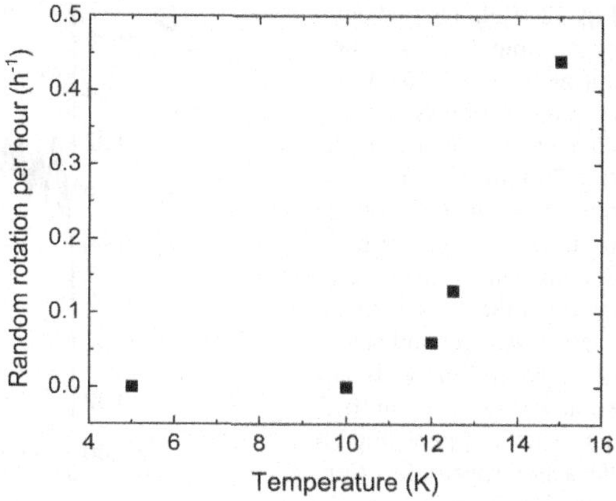

Fig. 11 Detecting random rotation over a long period of time by repeating STM images at non-disruptive parameters ($I = 5$ pA and $V = 10$ mV). Plot of rotational event per hour over stepwise increase of temperature

h in Fig. 11). In agreement with the micro-reversibility principle, this pure thermal excitation does not lead to any unidirectional rotation.

On the other hand, at $T = 5$ K we have also detected the first unidirectional rotation event by using low voltage bias pulses ($V < 370$ mV). We therefore conducted a series of measurements by gradually lowering the STM bias voltage to investigate how the time required to observe a one-step unidirectional rotation event increase, ranging from a few seconds to several hours. The experiment was systematically repeated on the same molecule following a reduction in bias voltage and an extension of the pulsing time duration, maintaining all the time a $I = 700$ pA constant tunnelling current. At 150 mV, the initial one-step directed rotation event was observed after keeping the STM tip apex above the molecule-rotor for 35,950 s (approximately 10 h) during a 12-h pulse. Importantly, there was negligible lateral thermal drift in our experimental setup. Each response time data point corresponds to a specific point on the z(t) curve, similar to the representation in Fig. 12a. The corresponding $Y = R(I/e)$ action spectrum was plotted for $n = 139$ experiments as presented in Fig. 12b.

For $V > 400$ mV, we confirmed a nearly consistent yield of approximately 5×10^{10} rotations per electron as discussed in a previous section. It was attributed to the activation of the C–H stretch vibrational mode at 370 meV. It is the principal energy entry port for initiating a given one-step rotation event in the higher bias voltage range, and it reveals an inelastic tunnelling event.

At lower bias voltages, we observed a sharp decline in yield between 330 and 400 mV, followed by the emergence of another nearly stable region in the 200–300 mV range. As the bias voltage dropped below 200 mV, we witnessed a rapid reduction in yield, eventually reaching a low yield of 6×10^{-15} rotations per electron at 150 mV. Through data fitting [38] (represented by the solid line in Fig. 12b), we identified a second onset at 165 meV (onset **i**) complementary to the 370 meV one (onset **ii**). These thresholds are associated with significant concentrations of DMNI-P deformation modes (specifically, the C–C stretch manifold) and the previously mentioned C–H stretch mode of DMNI-P, respectively. At low bias voltage, these intramolecular mechanical modes play a crucial role in triggering unidirectional

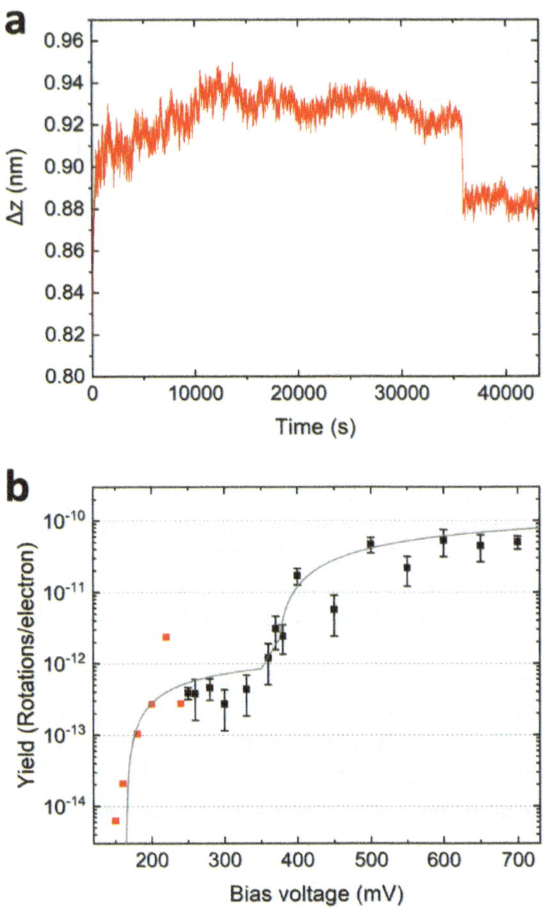

Fig. 12 Relative tip height versus time Δz(t) measured during the pulse. (**a**) A rotation event was observed after 35,950 s in a 12 h pulse (I = 700 pA, V = 150 mV). **b** Yield versus bias voltage for the rotor at pulses I = 700 pA. Due to the long waiting time, no statistical analysis was possible and only one (red-marked) data point was recorded in (**b**). The data have been fitted by the action spectra function of Ref. [38] for the two modes: 165 and 370 mV respectively

rotation. They provide energy to the collective rotation mode of the molecule. giving inelastic rotational energy, the other part being IVR redistributed and then release to the Au(111) surface.

Then, we have repeated the above rotation experiments by applying STM voltage pulses between 50 mV $\leq$ V $\leq$ 500 mV but also by increasing progressively the temperature from 5 K < T $\leq$ 17 K. Notice that this bias voltage range was selected according to the two inelastic onsets (165 meV and 370 meV) observed in Fig. 12. In these conditions (when T > 5 K), a faster rate of unidirectional rotation is observed using the voltage pulse technique. Depending on the voltage pulse setpoints (Fig. 13), the rate of a one-way rotation is plotted for three fixed temperatures (all data point measured at I = 700 pA). Increasing the temperature indeed leads to a faster rate, while this effect diminishes by increasing the voltages and eventually becomes weaker at V > 500 mV. The results indicate that at this energy, the awaited C–H vibrational mode (V > 370 mV) is excited. Here, we recover the inelastic electronic tunnel effect as the main energy entry port for the directed rotation. It overcomes the thermal contribution in the STM single molecule junction.

As summarized in Fig. 14, the combined thermal and electronic measurements have been extended towards all the experimental available temperature range. In Fig. 14, the blue dots represent the observed unidirectional rotation success while triggered by an STM voltage pulse resulting in an elastic or inelastic tunnelling current through the molecule. The orange dots indicate the occurrences of the random rotation events. The dashed lines serve to divide the plot into three distinct regions, the two right regions corresponding to the fitted yield for inelastic electronic excitation at T = 5 K (as shown in the Fig. 12b): they mark the excitation energy thresholds for the soft mechanical mode at 165 meV and the C–H stretch mode at 370 meV, respectively.

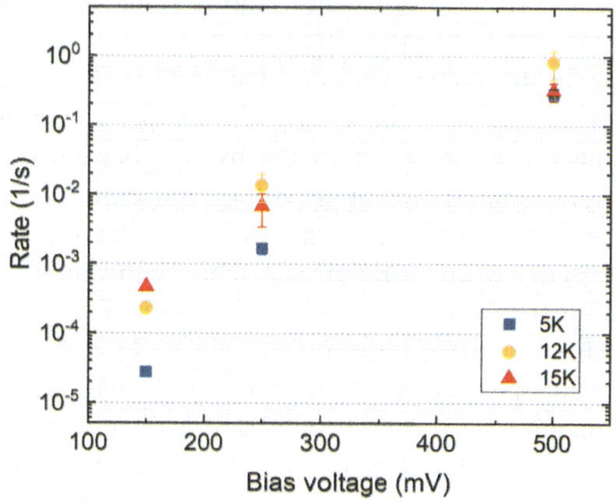

Fig. 13 Interplay between thermal and inelastic electronic excitations for unidirectional rotation. Rotational rate (1/s) dependence on the bias voltage at different temperatures (5, 12, and 15 K) at I = 700 pA

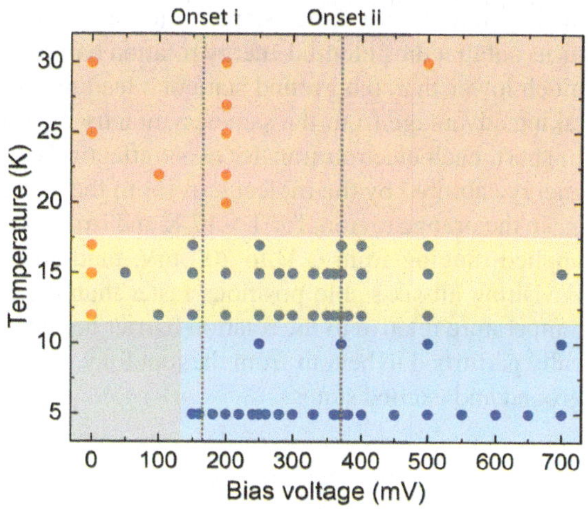

Fig. 14 Summary of combining thermal and electronic excitations. The orange dots indicate the conditions for random rotation and the blue ones the unidirectional rotations

In the grey region, no rotation (random or not random) was observed because of the extremely low temperature and bias voltages. In these conditions, the minute thermal energy is not enough to create a minimum energy path on the ground state potential energy surface of the molecule to overcome the above measured rotation energy barrier. The tunnelling current is purely elastic in those bias voltage conditions and provide no energy for this rotation.

In the light blue region, as discussed above and for 5 K < T < 12 K, STM voltage pulses from 150 mV < V < 700 mV can be used to induce stepwise CW or CCW rotation with 99% success rate. The inelastic tunnelling electrons provide the energy for the rotation and the IVR process. The molecule surface chirality is encrypted in the electronic structure of the molecule leading to an asymmetry between potential energy surface landscapes of ground and excited states, at least in the collective rotation direction. The micro reversibility principle can still be applied on the ground state PES. However the rotation process is not involved only in this ground state, thus explaining the observed one-way rotation.

In the yellow region, blue and orange dots are present. As the applied voltage goes near 0 V and within the temperature range of 12–20 K, molecules undergo random rotations activated by the thermal energy. More interestingly, within the same temperature range, voltage pulses in the observed range between 50 and 250 mV give rise to unidirectional rotation with an increased rate (as shown in Fig. 13) but in absence of inelastic effects. This is attributed to the origin of the tunnelling current (intensity measured in average external and away from the STM tunnel junction). The tunnelling current intensity results from millions of elastic electrons transfer events per second through the molecule. Each electron transfer event involves the quantum mixing of ground and excited virtual electronic configurations entangling left and right electronic reservoirs of the STM junction (the STM tip apex and the Au(111) metal surface with adsorbed molecule). Consequently, during each electron transfer event, part of the thermal energy can be redistributed on those excited states, thereby initiating a dynamic path on the excited state potential energy surface. It turns out that the potential energy rotation barrier along the excited state is in general much lower than the ground state one leading to a one-step unidirectional rotation taking advantage from the same asymmetry as discussed for the blue region above. In short, each electron transfer event effectively "projects" a portion of the thermal energy captured by the molecule-rotor to the excited states.

In the orange region, for T > 17 K and irrespective of the voltage pulse amplitude applied ranging from 0 V to 700 mV, random rotation persists and the molecule is visiting all six stable positions faster than the STM scanning rate. At such large temperature relative to the rotation barrier height, the molecule-rotor is too energetically perturbed to benefit from the topology of the potential energy landscape of the ground and excited states.

9 Conclusion

We have compared two DMBI-like molecular derivatives functioning like single molecule-rotors when chemisorbed on the Au(111) surface. In agreement with the micro-reversibility principle, both molecule-rotors demonstrate random rotation when only powered by a pure thermal excitation. In the electronic ground state of those molecules, the required thermal energy for an on-step random motion is higher for DMNIP due to the larger van der Waals interactions with the Au(111) surface. Both molecule-rotors can be driven one-way by STM voltage pulses. Interestingly and considering the extended number of phenyl rings of DMNI-P as compared to DMBI-P, both rotors exhibit the same vibrational onset energy entry port at V = 370 mV. We have also chosen the DMNI-P molecule for studying the interplay between thermal and elastic tunnelling. While the two distinguishable regimes of thermal and inelastic tunnelling excitations can be clearly separated, there exist an intermediate regime where the thermal contribution from the Au(111) surface enhances the rate for an unidirectional rotation of the molecule-rotor while maintaining an elastic tunnelling current through this molecule.

Acknowledgements This work was funded by the European Union. Views and opinions expressed are, however, those of the authors only and do not necessarily reflect those of the European Union. Neither the European Union nor the granting authority can be held responsible for them. This work has received funding from the European Innovation Council (EIC) under the project ESiM (grant agreement No 101046364). Support by the German Research Foundation (DFG) by the DFG Project 43234550, the Collaborative Research Centre (CRC) 1415 is gratefully acknowledged.

References

1. Stipe, B.C., Rezaei, M.A., Ho, W.: Inducing and viewing the rotational motion of a single molecule. Science **279**, 1907 (1998)
2. Gimzewski, J., Joachim, C., Schlittler, R.R., Langlais, V., Tang, H., Johannsen, I.: Rotation of a single molecule within a supramolecular bearing. Science **281**, 531 (1998)
3. Sainoo, Y., Kim, Y., Okawa, T., Komeda, T., Shigekawa, H., Kawai, M.: Excitation of molecular vibrational modes with inelastic scanning tunneling microscopy processes: examination through action spectra of cis-2-Butene on Pd(110). Phys. Rev. Lett. **95**, 246102 (2005)
4. Henningsen, N., Franke, K.J., Torrente, I.F., Schulze, G., Priewisch, B., Rück-Braun, K., Dokić, J., Klamroth, T., Saalfrank, P., Pascual, J.I.: Inducing the rotation of a single phenyl ring with tunneling electrons. J. Phys. Chem. C **111**, 14843 (2007)
5. Kudernac, T., Ruangsupapichat, N., Parschau, M., Maciá, B., Katsonis, N., Harutyunyan, S.R., Ernst, K.-H., Feringa, B.L.: Electrically driven directional motion of a four-wheeled molecule on a metal surface. Nature **479**, 208 (2011)
6. Tierney, H.L., Murphy, C.J., Jewell, A.D., Baber, A.E., Iski, E.V., Khodaverdian, H.Y., McGuire, A.F., Klebanov, N., Sykes, E.C.H.: Experimental demonstration of a single-molecule electric motor. Nat. Nanotechnol. **6**, 625 (2011)
7. Joachim, C., Rapenne, G.: Molecule concept nanocars: chassis, wheels, and motors? ACS Nano **7**, 11 (2013)
8. Erbas-Cakmak, S., Leigh, D.A., McTernan, C.T., Nussbaumer, A.L.: Artificial molecular machines. Chem. Rev. **115**, 10081 (2015)

9. Ohmann, R., Meyer, J., Nickel, A., Echeverria, J., Grisolia, M., Joachim, C., Moresco, F., Cuniberti, G.: Supramolecular rotor and translator at work: on-surface movement of single atoms. ACS Nano **9**, 8394 (2015)
10. Eisenhut, F., Durand, C., Moresco, F., Launay, J.-P., Joachim, C.: Training for the 1st international nano-car race: the Dresden molecule-vehicle. Eur. Phys. J. Appl. Phys. **76**, 10001 (2016)
11. Soe, W.-H., Shirai, Y., Durand, C., Yonamine, Y., Minami, K., Bouju, X., Kolmer, M., Ariga, K., Joachim, C., Nakanishi, W.: Conformation manipulation and motion of a double paddle molecule on an Au(111) surface. ACS Nano **11**, 10357 (2017)
12. Lorente, N., Joachim, C.: In: Joachim, C. (ed.) Rotations of Adsorbed Molecules Induced by Tunneling Electrons in Building and Probing Small for Mechanics, p. 181. Springer International Publishing, Cham (2020)
13. van Venrooy, A., García-López, V., Li, J.T., Tour, J.M., Dubrovskiy, A.V.: Nanocars with permanent dipoles: preparing for the second international nanocar race. J. Org. Chem. **85**, 13644 (2020)
14. Li, Z., Gao, C.Y., Su, D., Jia, C., Guo, X.: Principles of molecular machines at the single-molecule scale. ACS Mater. Lett. **3**, 1484 (2021)
15. Srivastava, G., Štacko, P., Mendieta-Moreno, J.I., Edalatmanesh, S., Kistemaker, J.C.M., Heideman, G.H., Zoppi, L., Parschau, M., Feringa, B.L., Ernst, K.-H.: Driving a third generation molecular motor with electrons across a surface. ACS Nano **17**, 3931 (2023)
16. Simpson, G.J., Persson, M., Grill, L.: Adsorbate motors for unidirectional translation and transport. Nature **621**, 82 (2023)
17. Nickel, A., Ohmann, R., Meyer, J., Grisolia, M., Joachim, C., Moresco, F., Cuniberti, G.: Moving nanostructures: pulse-induced positioning of supramolecular assemblies. ACS Nano **7**, 191 (2013)
18. Moresco, F., Meyer, G., Rieder, K.-H., Tang, H., Gourdon, A., Joachim, C.: Conformational changes of single molecules induced by scanning tunneling microscopy manipulation: a route to molecular switching. Phys. Rev. Lett. **86**, 672 (2001)
19. Kühne, T., Au-Yeung, K.H., Eisenhut, F., Aiboudi, O., Ryndyk, D.A., Cuniberti, G., Lissel, F., Moresco, F.: STM induced manipulation of Azulene-based molecules and nanostructures: the role of the dipole moment. Nanoscale **12**, 24471 (2020)
20. Soe, W.-H., Kleinwächter, M., Kammerer, C., Rapenne, G., Joachim, C.: Mechanics of molecule-gears with six long teeth. J. Phys. Chem. C **124**, 22625 (2020)
21. Lin, H.-H., Croy, A., Gutierrez, R., Cuniberti, G.: Atomistic Modelling of Energy Dissipation in Nanoscale Gears, p. 157. Springer International Publishing, Cham (2023)
22. Simpson, G.J., García-López, V., Petermeier, P., Grill, L., Tour, J.M.: How to build and race a fast nanocar. Nat. Nanotechnol. **12**, 604 (2017)
23. Ren, J., Freitag, M., Schwermann, C., Bakker, A., Amirjalayer, S., Rühling, A., Gao, H.-Y., Doltsinis, N.L., Glorius, F., Fuchs, H.: A unidirectional surface-anchored N-heterocyclic carbene rotor. Nano Lett. **20**, 5922 (2020)
24. Leff, H.S., Rex, A.F.: Resource letter MD-1: Maxwell's demon. Am. J. Phys. **58**, 201 (1990)
25. Launay, J.-P.: From Single Molecules to Practical Devices in From Non-Covalent Assemblies to Molecular Machines, pp. 379 (2010)
26. Burwell, R.L., Jr., Pearson, R.G.: The principle of microscopic reversibility. J. Phys. Chem. **70**, 300 (1966)
27. Dence, J.B.: Heat capacity and the equipartition theorem. J. Chem. Educ. **49**, 798 (1972)
28. Koumura, N., Zijlstra, R.W.J., van Delden, R.A., Harada, N., Feringa, B.L.: Light-driven monodirectional molecular rotor. Nature **401**, 152 (1999)
29. Kazaryan, A., Kistemaker, J.C.M., Schäfer, L.V., Browne, W.R., Feringa, B.L., Filatov, M.: Understanding the dynamics behind the photoisomerization of a light-driven fluorene molecular rotary motor. J. Phys. Chem. A **114**, 5058 (2010)
30. Gao, L., Liu, Q., Zhang, Y.Y., Jiang, N., Zhang, H.G., Cheng, Z.H., Qiu, W.F., Du, S.X., Liu, Y.Q., Hofer, W.A., Gao, H.J.: Constructing an array of anchored single-molecule rotors on gold surfaces. Phys. Rev. Lett. **101**, 197209 (2008)

31. Au Yeung, K.H., Kühne, T., Eisenhut, F., Moresco, F.: Anchoring Molecular Rotors by On-Surface Synthesis, p 117. Springer International Publishing, Cham (2020)
32. Zhang, Y., Kersell, H., Stefak, R., Echeverria, J., Iancu, V., Perera, U.G.E., Li, Y., Deshpande, A., Braun, K.F., Joachim, C., Rapenne, G., Hla, S.W.: Simultaneous and coordinated rotational switching of all molecular rotors in a network. Nat. Nanotechnol. **11**, 706 (2016)
33. Wei, P., Menke, T., Naab, B.D., Leo, K., Riede, M., Bao, Z.: 2-(2-Methoxyphenyl)-1,3-dimethyl-1H-benzoimidazol-3-ium Iodide as a new air-stable n-type dopant for vacuum-processed organic semiconductor thin films. J. Am. Chem. Soc. **134**, 3999 (2012)
34. Lüssem, B., Keum, C.-M., Kasemann, D., Naab, B., Bao, Z., Leo, K.: Doped organic transistors. Chem. Rev. **116**, 13714 (2016)
35. Schwarze, M., Naab, B.D., Tietze, M.L., Scholz, R., Pahner, P., Bussolotti, F., Kera, S., Kasemann, D., Bao, Z., Leo, K.: Analyzing the n-doping mechanism of an air-stable small-molecule precursor. ACS Appl. Mater. Interfaces **10**, 1340 (2018)
36. Eisenhut, F., Kühne, T., Monsalve, J., Srivastava, S., Ryndyk, D.A., Cuniberti, G., Aiboudi, O., Lissel, F., Zobač, V., Robles, R., Lorente, N., Joachim, C., Moresco, F.: One-way rotation of a chemically anchored single molecule-rotor. Nanoscale **13**, 16077 (2021)
37. Au-Yeung, K.H., Sarkar, S., Kühne, T., Aiboudi, O., Ryndyk, D.A., Robles, R., Lorente, N., Lissel, F., Joachim, C., Moresco, F.: A nanocar and rotor in one molecule. ACS Nano **17**, 3128 (2023)
38. Kim, Y., Motobayashi, K., Frederiksen, T., Ueba, H., Kawai, M.: Action spectroscopy for single-molecule reactions–experiments and theory. Prog. Surf. Sci. **90**, 85 (2015)
39. Kawai, M., Komeda, T., Kim, Y., Sainoo, Y., Katano, S.: Single-molecule reactions and spectroscopy via vibrational excitation. In: Philosophical Transactions of the Royal Society A: Mathematical, Physical and Engineering Sciences, p 1163 (2004)
40. Au-Yeung, K.H., Sarkar, S., Kühne, T., Aiboudi, O., Ryndyk, D.A., Robles, R., Lissel, F., Lorente, N., Joachim, C., Moresco, F.: Thermal with electronic excitation for the unidirectional rotation of a molecule on a surface. J. Phys. Chem. C **127**, 16989 (2023)

SPRINGER NATURE

GPSR Compliance

The European Union's (EU) General Product Safety Regulation (GPSR) is a set of rules that requires consumer products to be safe and our obligations to ensure this.

If you have any concerns about our products, you can contact us on ProductSafety@springernature.com

In case Publisher is established outside the EU, the EU authorized representative is:

Springer Nature Customer Service Center GmbH
Europaplatz 3
69115 Heidelberg, Germany